Pericyclic Reactions

SECOND EDITION

Ian Fleming

T0130859

OXFORD

UNIVERSITY PRESS

OXFORD

UNIVERSITY PRESS

Great Clarendon Street, Oxford, OX2 6DP,
United Kingdom

Oxford University Press is a department of the University of Oxford.
It furthers the University's objective of excellence in research, scholarship,
and education by publishing worldwide. Oxford is a registered trade mark of
Oxford University Press in the UK and in certain other countries

First Edition 1998

Published in the United States of America by Oxford University Press
198 Madison Avenue, New York, NY 10016, United States of America

British Library Cataloguing in Publication Data
Data available
Library of Congress Control Number: 2014948247

ISBN 978-0-19-968090-0

Printed in Great Britain by
Ashford Colour Press Ltd, Gosport, Hampshire

Preface

Pericyclic chemistry is over 50 years old now, and its history is no longer a necessary part of the subject. Accordingly, I have barely commented in the text on how the subject developed, so perhaps I can say something here to place it in a historical context.

By the end of the 1950s the main features of ionic and radical reactions were reasonably well understood, but pericyclic reactions were not even recognized as a separate class. Diels–Alder reactions, and a good many others, were known individually. Curly arrows were used to show where the bonds went to in these reactions, but the absence of a sense of direction to the arrows was unsettling. Doering, with tongue in cheek, even called some of them 'no-mechanism reactions' in the early 1960s. At that time, there were a number of such reactions, often with disconcerting aspects, which were simply not understood because they seemed to fly in the face of all sense. Why, if dienes added to alkenes, didn't alkenes easily dimerize to make cyclobutanes—the reaction was known to be exothermic but rarely happened? On the other hand, acetylenes and allenes did give four-membered rings when treated with electrophiles—why were they different? Why did cis-3,4-dimethylcyclobutene open to give the thermodynamically less stable cis,trans-butadiene? Why did calciferol transfer a hydrogen atom easily from C-9 to C-18 (steroid numbering; it is a shift of seven atoms along a triene), whereas most trienes, like cycloheptatriene, where it ought to have been even easier, didn't? Why did pentadienyl cations cyclize, with bond formation from C-1 to C-5, when both atoms, since they both carried partial positive charge, ought to repel each other?

Most organic chemists at the time didn't notice these problems, and those who did were not aware of how many there were or how much they resembled each other. This began to change in the autumn of 1963, when R. B. Woodward, in the course of his synthesis of vitamin B_{12}, came across another seemingly inexplicable example of counter-thermodynamic stereochemistry—it would now be recognized as the disrotatory electrocyclic ring closing of a hexatriene. The reactions I listed as questions above were in his mind, and now, with his very own reaction to think about as well, he recognized a pattern, and sought an explanation for all these bewildering observations in molecular orbital theory. He and Roald Hoffmann, starting in 1965, introduced a series of rules governing the stereochemistry of the various classes of pericyclic reactions. They showed, with a significant contribution from Abrahamson and Longuet-Higgins, how to explain them all using correlation diagrams based on the symmetry of the molecular orbitals. Then in 1969 they introduced the word *pericyclic*, and showed that all the rules could be subsumed in the pair of rules used in this book.

They were exciting times, as order was revealed, as explanation followed, and as predictions were made and fulfilled. I was there when Ranganathan first brought the B_{12} result to light, and heard Woodward at group meetings discussing the reactions I listed as questions here. I will finish with a very personal memory of his drawing for me, at my bench late one evening in the spring of 1964, the four atoms of butadiene and the six atoms of hexatriene, with the lobes of the p-orbitals filled in for what we would now call the HOMO, relating the pattern that emerged to Ranganathan's results, and saying, 'It must have something to do with the molecular orbitals.' From this tentative perception, the subject grew to maturity in six intense years. I hope this book will bring you to an appreciation of the beautiful pattern that emerged, to become one of the fundamental branches of organic chemistry.

Ian Fleming
Cambridge
August 2014

Table of Contents

The nature of pericyclic reactions

1.1 Introduction

In contrast to the common ionic and radical reactions of organic chemistry, pericyclic reactions are a third distinct class. They have cyclic transition structures in which all bond-forming and bond-breaking takes place in concert, without the formation of an intermediate.

1.2 Ionic, radical, and pericyclic reactions

Every organic reaction can be classified into one of three, more-or-less exclusive categories—ionic, radical, and pericyclic. Ionic reactions involve pairs of electrons moving in one direction. In a unimolecular reaction, like the ionization of a tertiary alkyl halide, the carbon–halogen bond cleaves with both electrons going to the chloride ion, leaving an electron deficiency behind on the carbocation. In a bimolecular ionic reaction, one component, called the nucleophile, provides both electrons for a new bond, and the other, called the electrophile, receives them, as in the aldol reaction between the enolate of acetone and acetone itself. Ionic reactions are by far the most numerous of organic reactions.

a breaking bond,
with both electrons moving to
one component

a new bond,
with both electrons supplied by
the nucleophile

In ionic reactions, curly arrows have two functions: they identify where the electrons come from and where they are going, and hence identify which component is the nucleophile and which the electrophile; simultaneously they identify which bonds are broken and which new bonds are made. If the tail of an arrow is from a bond, then that bond is breaking, and if the head of the arrow falls between two atoms it shows where the new bond is being made.

Radical reactions involve the correlated movement of single electrons. In a unimolecular reaction like the photolytic cleavage of chlorine, one electron moves to one atom and the other electron moves to the other. In bimolecular reactions, a new bond is formed when one electron from one component pairs up with an electron from the other component, as with the removal of a hydrogen atom from toluene by a chlorine atom to give the benzyl radical and hydrogen chloride. This type of reaction is illustrated with fishhook arrows. A useful short cut when illustrating radical reactions is to draw only the alternate arrows, since each implies the existence of its partner in the making or breaking of a bond; but the full complement of fishhook arrows is used here, in order to emphasize the symmetrical separation in the breaking of the Cl-Cl bond, and the equal contribution of electrons from each component in the making of the H-Cl bond, in contrast to ionic reactions, where there is a definite sense of direction to the movement of the electrons.

Fishhook arrows have the same functions of identifying simultaneously where the electrons come from and go to, and which bonds have been broken and which have been made. However, since each component provides one electron for each new bond, there is strictly no nucleophile and electrophile.

the breaking bond, with one electron moving onto one atom, and the other onto the other atom

the new bond, with the electrons supplied equally by both components

Pericyclic reactions are the third distinct class. They have cyclic transition structures in which all bond-forming and bond-breaking takes place in concert, without the formation of an intermediate. The Diels–Alder reaction between a *diene*, like butadiene **1.1**, and a *dienophile*, like maleic anhydride **1.2**, to give a *cycloadduct* **1.3** is by far the most important. Related to it is the Alder 'ene' reaction between an *alkene*, propene **1.4**, and an *enophile*, maleic anhydride again, to give the anhydride **1.5**. The curly arrows can be drawn to illustrate where new bonds are formed in either direction—clockwise, as here, but equally well anti-clockwise. They could even be drawn with fishhook arrows, and would still describe the same reaction. There is no absolute sense in which the electrons flow from one component to the other. Similarly, there is no absolute sense in which the hydrogen atom that moves from one carbon atom to the other in the ene reaction is a hydride shift, as seems to be implied by the curly arrows here, or a proton shift, as it would seem to be if the arrows were to have been drawn in the opposite direction. In other words, neither component can be associated with the supply of electrons to any of the new bonds. The curly arrows, therefore, have a somewhat different meaning from those used in ionic reactions.

The curly arrows in a pericyclic reaction share the capacity that they have in ionic reactions to show which bonds are breaking and where new bonds are forming, but they do not show the direction of electron flow.

1.1   1.2   1.3   1.4   1.2   1.5

A Diels-Alder reaction   An Alder ene reaction

In this they somewhat resemble the curly arrows used to show resonance in benzene where the arrows show where to draw the new bonds and which bonds not to draw in the canonical structure, but in this case there is neither a sense of direction nor even an actual movement. The analogy between the resonance of benzene and the electron shift in the Diels–Alder reaction is not far-fetched, but it is as well to be clear that one is a reaction, with starting materials and a product, and the other is not.

The double-headed arrow showing the relationship between canonical structures should only be used for resonance, and never for an equilibrium reaction.

1.3 The four classes of pericyclic reactions

All pericyclic reactions share the feature of having a cyclic transition structure, with a concerted movement of electrons simultaneously breaking bonds and making bonds. Within that overall category, it is convenient to divide pericyclic reactions into four main classes. These are *cycloadditions, electrocyclic reactions, sigmatropic rearrangements*, and *group transfer reactions*, each of which possesses special features not shared by the others, and some of which employ a terminology that cannot be used without confusion in the other classes. It is a good idea to be clear which class of reactions you are talking about, in order to avoid using inappropriate terminology. We shall begin by illustrating the identifying features of each of the classes in this chapter. Cycloadditions are the largest class. The range of reactions, the stereochemistry they show, the regioselectivity when unsymmetrical components are used, and the patterns of which cycloadditions occur and which do not–these are all discussed in Chapter 2. The theoretical basis for the patterns of reactivity is then explained in Chapter 3. We return to the other three classes of pericyclic reaction in Chapters 4-6, armed, by then, with a theoretical framework with which to discuss them. For now we shall look at them all briefly, simply to establish what they are.

The four classes of pericyclic reaction:

Cycloadditions
Electrocyclic reactions
Sigmatropic rearrangements
Group transfer reactions

1.4 Cycloadditions

Cycloadditions are characterized by two components' coming together to form two new σ-bonds, at the ends of both components, joining them together to form a ring, with a reduction in the length of the conjugated system of orbitals in each component. Cycloadditions are by far the most abundant, featureful, and useful of all pericyclic reactions. They are inherently bimolecular, although they can be unimolecular when the two components are connected by a chain of atoms.

Diels–Alder reactions like that of butadiene **1.1** with maleic anhydride **1.2** are cycloadditions mobilizing six electrons (three curly arrows). The dimerization of cyclopentadiene **1.6** is another Diels–Alder reaction, but it also illustrates the inherent reversibility of any cycloaddition–cracking the dimer **1.7** on heating is called a retro-cycloaddition or a **cycloreversion**.

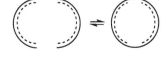

cycloaddition

cycloreversion

1.6   **1.6**   **1.7**

1,3-Dipolar cycloadditions are another important family, with the impressive sequence of reactions involved when ozone reacts with an alkene as an example here. At −78°, ozone adds **1.8** (arrows) to give the molozonide **1.9**. On warming, this undergoes a 1,3-dipolar cycloreversion (**1.9** arrows), and then another 1,3-dipolar cycloaddition (**1.10 + 1.11**, arrows) with the opposite regiochemistry, to give the ozonide **1.12**, which is normally the only product that can be isolated.

The Criegee mechanism for ozonolysis: a dramatic sequence with successively a 1,3-dipolar cycloaddition, a 1,3-dipolar cycloreversion, and another 1,3-dipolar cycloaddition, all taking place below room temperature.

Cheletropic reactions are a special group of cycloadditions or cycloreversions in which two σ bonds are made or broken to the *same* atom. Thus, sulfur dioxide adds to butadiene **1.1** to give an adduct **1.13** for which the sulfur has provided a lone pair to one of the σ bonds and has received electrons in the formation of the other. Concomitantly, the sulfur has changed from S(IV) to S(VI). The reaction is readily reversible on heating, and so sulfur dioxide can be used to protect dienes, and likewise the adduct **1.13** can be heated in the presence of dienophiles as a convenient (liquid) source of butadiene.

The distinguishing feature of a cheletropic cycloaddition

both σ-bonds made to the same atom

1.5 Electrocyclic reactions

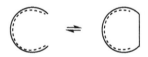

Whereas cycloadditions are bimolecular, with two components coming together to form two new σ-bonds, **electrocyclic reactions** are invariably unimolecular. They are characterized by the creation of a ring from an open-chain conjugated system, with a σ-bond forming across the ends of the conjugated system, and with the conjugated system becoming shorter by one p-orbital at each end.

Representative reactions in this class are the ring-opening of cyclobutene **1.14** on heating to give butadiene **1.1**, and the ring-closing of hexatriene **1.15** to give cyclohexadiene **1.16**. Notice that the reactions are inherently reversible, and that the directions they take are determined by thermodynamics. Most electrocyclic reactions are ring-closings, since a σ-bond is created at the expense of a π-bond, but a few are ring-openings, because of the relief of the ring strain in a small ring.

1.14   **1.1**   **1.15**   **1.16**

An electrocyclic ring-opening and an electrocyclic ring-closing

1.6 Sigmatropic rearrangements

Sigmatropic rearrangements are unimolecular isomerizations. They are characterized by the movement of a σ-bond from one position to another, with a concomitant movement of the conjugated systems to accommodate the new bond and fill in the vacancy left behind. The oldest known example **1.17 → 1.18** is the first step in the Claisen rearrangement, when a phenyl allyl ether is heated. What makes it a sigmatropic rearrangement is the movement of the single bond in **1.17**, drawn in bold, to its new position in **1.18**. It has effectively moved three atoms along the carbon chain (from C-1 to C-3), and three atoms along the chain of the oxygen atom and two carbon atoms (O-1′ to C-3′). This type of rearrangement is called a [3,3] shift, with the numbers identifying the number of atoms along the chain that each end of the bond has moved. The second step, forming the phenol **1.19**, is an ordinary ionic reaction—the enolization of a ketone. It is perhaps a timely reminder that ionic reactions often precede or follow a pericyclic reaction, sometimes disguising the pericyclic event.

A [3,3] sigmatropic rearrangement

1.17   **1.18**   **1.19**

A quite different-looking rearrangement is the hydride shift **1.20 → 1.21**, also long known from the chemistry of vitamin D. In this case, the end of the H–C bond attached to the hydrogen atom (H-1′) remains attached to the hydrogen, but the other end moves seven atoms (C-1 to C-7) along the conjugated carbon chain. This reaction is therefore called a [1,7] shift.

A [1,7] sigmatropic rearrangement

1.20   **1.21**

Another quite different-looking sigmatropic reaction is the Mislow rearrangement **1.22** → **1.23**, which is invisible because thermodynamically unfavourable; but the ease with which it takes place explains why allyl sulfoxides racemize so much more readily than other sulfoxides. Here, one end of the C–S bond moves from the sulfur (S-1′) to the oxygen atom (O-2′) and the other end moves from C-1 to C-3. This is therefore called a [2,3] shift, the bond marked in bold moving two atoms at one end and three at the other.

A [2,3] sigmatropic rearrangement

1.22   **1.23**

1.7 Group transfer reactions

There are only a few reactions in the class known as **group transfer reactions**. Ene reactions like that in Section 1.2 on p. 2 are the most common. Stripped to their essence, ene reactions have the form **1.4** + **1.24** → **1.25**. They usually take place from left to right, since overall a π-bond is replaced by a σ-bond, but they are, of course, reversible. They resemble [1,5] sigmatropic rearrangements, since a σ-bond moves, and they also resemble cycloadditions like Diels–Alder reactions, with one of the π-bonds of the diene replaced by a σ-bond. Nevertheless, since the reaction is bimolecular and no ring is formed, they are neither sigmatropic rearrangements nor cycloadditions.

1.4 1.24   1.25

Ene reactions have a hydrogen atom moving from the ene **1.4** to the enophile **1.24**, but other atoms can, in principle, move. In practice, the only elements other than hydrogen commonly employed in this kind of reaction are metals like lithium, magnesium, or palladium, when the reaction **1.26** is called a metalla-ene reaction.

A metalla-ene reaction

1.26

The carbon chain may also have one or more oxygen or nitrogen atoms in place of the carbons. Thus, if the atom carrying the hydrogen is an oxygen atom and the atom to which it is moving is also an oxygen atom, it becomes an aldol reaction. Such reactions are usually carried out with acid- or base-catalysis, and are rarely pericyclic in nature; but one reaction that probably is pericyclic is the thermal decarboxylation of a β-keto acid **1.27**, which also demonstrates that the fundamental reaction can take place from right to left.

A fragmentation that is also a retro-group-transfer reaction

1.27

The other well-known group transfer reaction is the concerted *syn* delivery of two hydrogen atoms from the reactive intermediate diimide **1.28** to an alkene or alkyne, driven by the formation of the stable molecule nitrogen.

1.28

1.8 Further reading

Most textbooks of organic chemistry include sections introducing pericyclic reactions, and there are dozens of slide shows and lecture notes available on-line. An extended version of the material in this book can be found in Chapter 6 in Fleming, I., Student Edn., 2009, Reference Edn. 2010, *Molecular Orbitals and Organic Chemical Reactions*, Chichester: Wiley.

Other books:

Sankaram, S., 2005, *Pericyclic Reactions—A Textbook: Reactions, Applications and Theory*, Chichester; Wiley.

On-line sites with animations and other visuals:

http://personalpages.manchester.ac.uk/staff/T.Wallace/StartPR.htm

http://www.chemtube3d.com/pericyclicintro.html

https://itunes.apple.com/us/course/pericyclic-reactions-in-organic/id562191620

1.9 Exercises

Each of the following transformations is the result of two successive pericyclic reactions. Draw the structures of the intermediates A–E, and identify the class of pericyclic reaction to which each step belongs:

Cycloaddition reactions

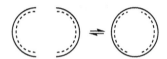

2.1 Introduction

Cycloadditions are the most useful of all pericyclic reactions in organic synthesis. This chapter describes the wide range of known cycloadditions, identifies the conditions under which they take place, draws attention to their regio- and stereochemistry, and gives the simple rules for which of them take place and which do not. The explanations for most of these features, based on the molecular orbitals involved, will then be covered in the next chapter.

2.2 Diels–Alder reactions

The most important type of cycloaddition is the Diels–Alder reaction. Stripped down to its essential components, it is the reaction between butadiene **2.1** and ethylene **2.2** to form cyclohexene **2.3**. The ethylenic component is called the dienophile. In practice this reaction is unbearably slow. It can be forced to take place: after 17 hours at 165° and 900 atmospheres, it does give a yield of 78%; but it is advisable, for reasons we shall see in Chapter 3, for the dienophile to have attached to it one or more electron-withdrawing groups like a carbonyl, nitrile, nitro, or sulfonyl group. Thus, in the reaction with maleic anhydride illustrated in Chapter 1 on p. 2, the presence of the two carbonyl groups allows the reaction to take place at 20° in 24 h in quantitative yield. The presence of a carbonyl group does not change the core *components* of the reaction, which are the three double bonds affected by the three curly arrows. The carbonyl group is a substituent attached to the periphery, affecting the rate, but not changing the fundamental nature of the reaction. Substituents on the diene also increase the rate, provided they are electron-donating, like methyl, alkoxy or amino groups. Figs. 2.1 and 2.2 illustrate these points, showing the times and temperatures that have been used in a representative range of Diels–Alder reactions, and giving some sense of which ones are easy and which need a bit of persuasion before they can be used in synthesis.

The conjugated system on the four carbon atoms coming from the diene changes from four conjugated p-orbitals in the starting material to two in the product, and the conjugated system of the dienophile changes from two conjugated p-orbitals to none.

2.1 2.2   **2.3**

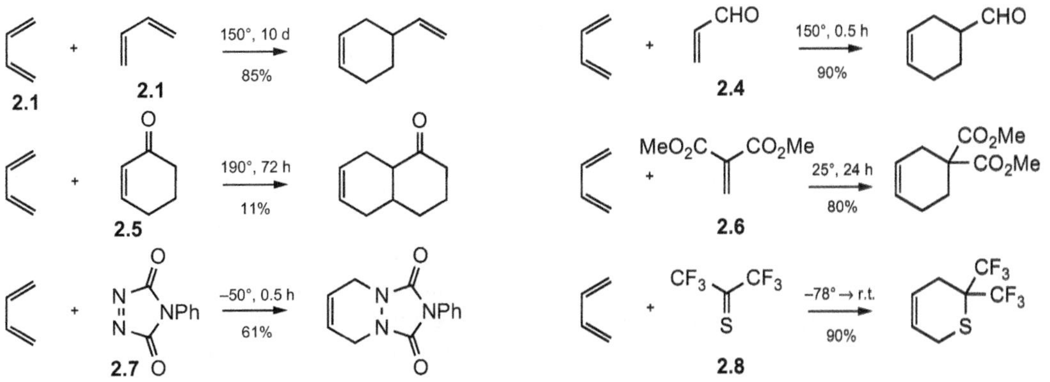

Fig. 2.1 Diels–Alder reactions of butadiene 2.1 with a range of representative dienophiles

Thus, if you look at the different dienophiles in Fig. 2.1, the dimerization of butadiene **2.1** is slower than its reaction with acrolein **2.4**. Methyl acrylate and methyl vinyl ketone have electron-withdrawing substituents Z of comparable power, and react at a similar rate, but cyclohexenone **2.5**, which has a β-alkyl substituent, is considerably less reactive. Nitroethylene has one of the most powerful electron-withdrawing groups, and is a very good dienophile, as is any alkene with two electron-withdrawing groups, like the methylenemalonate **2.6**, benzoquinone, or the maleic anhydride we saw earlier. An acetylenic ketone is a little more reactive than its ethylenic counterpart, and benzyne is exceptionally powerful in spite of its not having an electron-withdrawing group. The weak π-bonds in the triazolinedione **2.7** and the thioketone **2.8** make them especially good dienophiles; the former is one of the most reactive dienophiles that can be kept in a bottle.

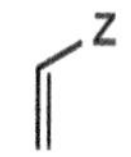

When we come to consider the effect of having a substituent on the diene, we see from the examples in Fig. 2.2 that an electron-donating substituent X makes dienes more reactive than butadiene. Even a group as weakly electron-donating as the methyl group in piperylene **2.9** and in isoprene **2.10** enables the reaction

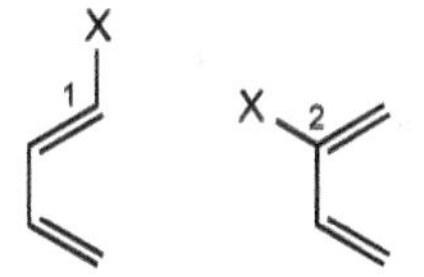

Z = COR, CN, NO$_2$, SO$_2$R etc.

X = Me, OMe, NMe$_2$ etc.

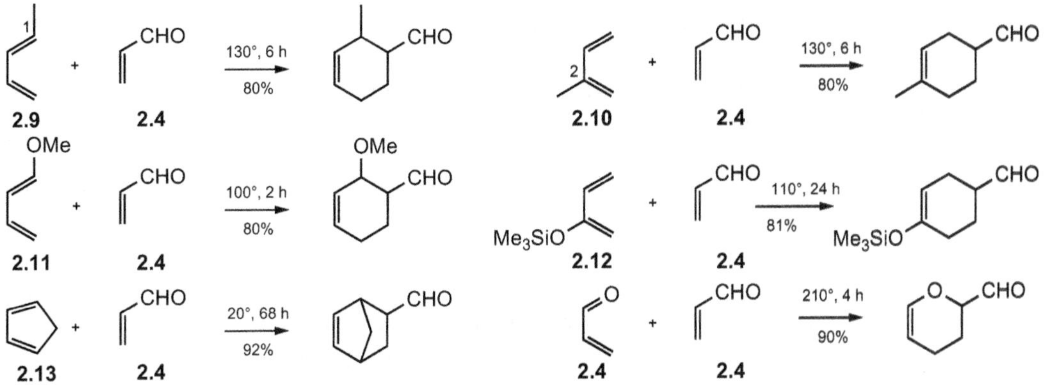

Fig. 2.2 Diels-Alder reactions of acrolein 2.4 with a range of representative dienes

to proceed at a lower temperature, whether the substituent is on C-1 or on C-2. A more powerful electron-donating group, like the methoxy group on C-1 in the diene **2.11** and the silyloxy group on C-2 in the diene **2.12**, makes these dienes even more reactive.

Whenever both the diene and the dienophile are unsymmetrically substituted, there are two possible isomeric adducts, which we describe as **regioisomers**. They are informally described as 'ortho', 'meta', and 'para'. Fortunately, in the energetically most favourable case, when the diene carries an electron-donating group X and the dienophile carries an electron-withdrawing group Z, there is a high degree of regioselectivity in favour of the 'ortho' and 'para' isomers. Thus, the regioisomers illustrated in Fig. 2.2 are substantially the major products in each case. We return to the topic of regioselectivity in Section 2.9 on p. 25.

'ortho' **major**   'meta' **minor**

'para' **major**   'meta' **minor**

s-*cis*

1 | 100

s-*trans*

The letter s- (for single bond) identifies this as a description of conformation, not configuration.

An impressive result in Fig. 2.2 is the evidence that the cyclic diene cyclopentadiene **2.13** is significantly more reactive than the open-chain dienes. A diene can participate in a Diels–Alder reaction only when it is in the s-*cis* conformation. If it were to react in the s-*trans* conformation, a *trans* double bond would have to be formed in the cyclohexene product, and a *trans* double bond in a cyclohexene is impossibly high in energy. The s-*trans* conformation in a diene is lower in energy than the s-*cis*, with butadiene itself having only about 1% of its molecules at room temperature in the s-*cis* conformation. Cyclic dienes are therefore more reactive, all other things being equal, because they have all their molecules in the s-*cis* arrangement, and do not have to add the difference in energy between the s-*trans* and the s-*cis* to the activation energy for the reaction. This is especially true of cyclopentadiene **2.13**. Butadienes with a C-2 substituent also spend more of their time in the s-*cis* conformation, which further increases the rates of Diels–Alder reactions with dienes like isoprene.

Figs. 2.1 and 2.2 also show, with the dienophiles **2.7** and **2.8**, and the diene **2.4**, that neither component is obliged to have an all-carbon skeleton. When dienes or dienophiles have a heteroatom like nitrogen, oxygen, or sulfur in the conjugated system, they are called **heterodienes** or **heterodienophiles**, and the reactions are called **hetero-Diels–Alder reactions**.

It is possible to have the electron-donating substituents X on the dienophile and the electron-withdrawing substituents Z on the diene; such reactions are said to have **inverse electron demand**. They are much less common, because having the substituents this way round is not as effective at increasing the rate of reaction as having them the usual way round.

Diels–Alder reactions are, of course, reversible, and the pathway followed for the reverse reaction (**2.3** arrows) can sometimes be as telling as the pathway for the forward reaction. The direction in which any pericyclic reaction takes place is determined by thermodynamics, with cycloadditions, like the Diels–Alder reaction, usually taking place to *form* a ring because two π-bonds on the left are replaced by two σ-bonds on the right. A Diels–Alder reaction can be made to take place in reverse when the products do not react with each other rapidly, as in the pyrolysis of cyclohexene **2.3** at 600°, when the two products are gases that separate. It helps if either the diene or the dienophile has some special stabilization not present in the starting adduct, as in the formation of the aromatic ring in anthracene **2.15** in the synthesis of diimide **2.16** from the adduct **2.14**, and in the extrusion of the stable molecule nitrogen **2.20** from the unstable intermediate **2.18**. A retro-cycloaddition can also show up when one of the products is consumed, dragging the equilibrium over, as in both examples below, where the reactive intermediates, diimide **2.16** and the o-quinodimethane **2.19**, are consumed in further reactions.

3 π-bonds ⟶ 1 π-bond
4 σ-bonds   6 σ-bonds

| 2.14 | 2.15 | 2.16 |

| 2.17 | 2.18 | 2.19 | 2.20 |

2.3 1,3-Dipolar cycloadditions

1,3-Dipoles react with alkenes or alkynes, or with heteroatom-containing double and triple bonds like carbonyl groups or nitriles, to form heterocyclic rings. The kind of dipoles that feature in 1,3-dipolar cycloadditions are isoelectronic with an allyl anion, having a conjugated system of three p-orbitals on three atoms, X, Y, and Z, and four electrons in the π-conjugated system **2.21**. The range of possible structures is large; X, Y, and Z are commonly almost any combination of C, N, O, and S, with a double or, in those combinations that can support it, a triple bond **2.22** between two of them. Likewise, the **dipolarophiles**, analogous to dienophiles, have a double or triple bond **2.23** between any pair A and B of the same common elements.

| 2.21 | 2.22 |

$A \equiv B$

2.23

2.21   **2.24**   **2.22**

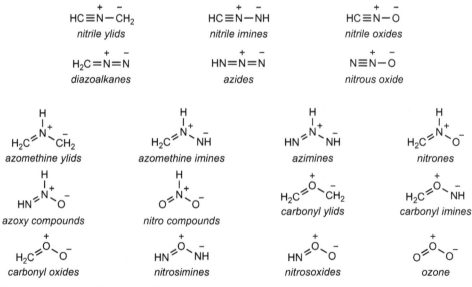

2.23   **2.25**   **2.23**   **2.26**

Ozone is a symmetrical 1,3-dipole with both ends equally 'nucleophilic' and equally 'electrophilic'.

Each canonical structure, say **2.21**, makes one end of the dipole appear to be a nucleophile, Z as drawn in **2.21**, and the other X to be an electrophile, but it is necessary only to draw the canonical structure **2.24** to reveal that the nucleophilicity and electrophilicity appear to have changed ends. Both ends have nucleophilic properties, and both have electrophilic properties; it is never safe to assume that one end is more nucleophilic just because it has been drawn with the formal negative charge.

Fig. 2.3 shows the core structures of the most important 1,3-dipoles and what they are all called. As with dienes, they can have electron-donating or withdrawing substituents attached at any of the atoms with a hydrogen atom in the core structure, and these modify the reactivity and selectivity that the dipoles show for different dipolarophiles. Some of the dipoles are stable enough to be isolated. These include ozone and diazomethane, and suitably substituted azides, nitrones, and nitrile oxides. Others, like the ylids, imines, and carbonyl oxides, are reactive intermediates that have to be made *in situ*. Fig. 2.4 shows some examples of some common 1,3-dipolar cycloadditions.

Several significant features of 1,3-dipolar cycloadditions are illustrated in Fig. 2.4. While diazopropane reacts with methyl methacrylate **2.27** straightforwardly to give the adduct **2.28**, the reaction with methyl acrylate **2.29** gives an

$$HC\equiv\overset{+}{N}-\overset{-}{CH_2}$$
nitrile ylids

$$HC\equiv\overset{+}{N}-\overset{-}{NH}$$
nitrile imines

$$HC\equiv\overset{+}{N}-\overset{-}{O}$$
nitrile oxides

$$H_2C=\overset{+}{N}=\overset{-}{N}$$
diazoalkanes

$$HN=\overset{+}{N}=\overset{-}{N}$$
azides

$$N\equiv\overset{+}{N}-\overset{-}{O}$$
nitrous oxide

azomethine ylids

azomethine imines

azimines

nitrones

azoxy compounds

nitro compounds

carbonyl ylids

carbonyl imines

carbonyl oxides

nitrosimines

nitrosoxides

ozone

Fig. 2.3 The names of the common 1,3-dipoles

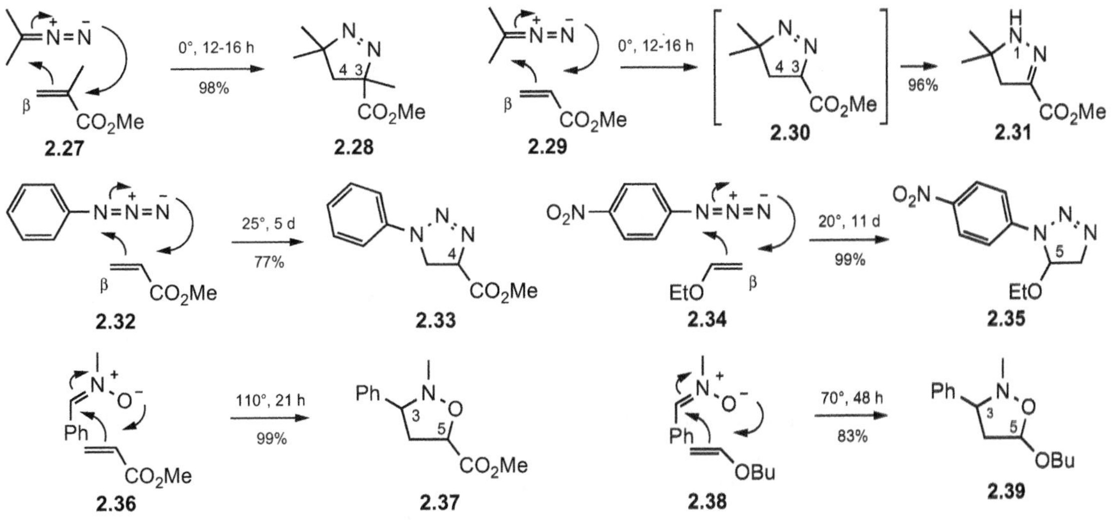

Fig. 2.4 Some representative 1,3-dipolar cycloadditions

adduct **2.30** that immediately loses the proton adjacent to the carbonyl group at C-3 and regains it at N-1 to give a tautomer **2.31**, which is the product actually isolated. With unsymmetrical dipoles and dipolarophiles, the reactions can always give two regioisomers. This is illustrated by the same pair of reactions, where the adducts **2.28** and **2.30** have the ester substituent cleanly at C-3 of the pyrazoline products, and there is no trace of the regioisomer with the ester group at C-4. The α,β-unsaturated esters are unmistakably electrophilic at the β-position; evidently the diazoalkane is more nucleophilic at the carbon end than at the nitrogen end of the dipole, in spite of the way the curly arrows are drawn. It is important to remember that there is no sense of direction to the arrows—they could equally be drawn the other way round using a canonical structure **2.40** of the diazoalkane.

With aryl azides, the reactions with alkenes that have an electron-withdrawing substituent **2.32** or an electron-donating substituent **2.34** show opposite regioselectivity. The nitrogen attached to the benzene ring evidently has nucleophilic character, bonding to the β-carbon of methyl acrylate, and the nitrogen at the terminus of the dipole has electrophilic character, bonding to the β-carbon of the ethyl vinyl ether. But a self-consistent picture like this is not carried over to all 1,3-dipoles. Nitrones, for example, react with methyl acrylate **2.36** and with a vinyl ether **2.38** with the same regiochemistry, with the substituent turning up on C-5 whether it is an acceptor or a donor.

Fig. 2.5 illustrates two of the many ways in which unstable dipoles, those that are only ever reactive intermediates, can be prepared and allowed to react *in situ*. Thus, the nitrile oxide **2.41** reacts with methyl acrylate **2.42** to give the adduct **2.43** with the ester group on C-5 of an isoxazoline ring. Nitrile oxides also react with simple terminal alkenes, and with the same regiochemistry, to place the alkyl group on C-5. In contrast, many dipoles react well only with electron-rich

The arrows in **2.27** could have been drawn anti-clockwise and would still mean the same reaction:

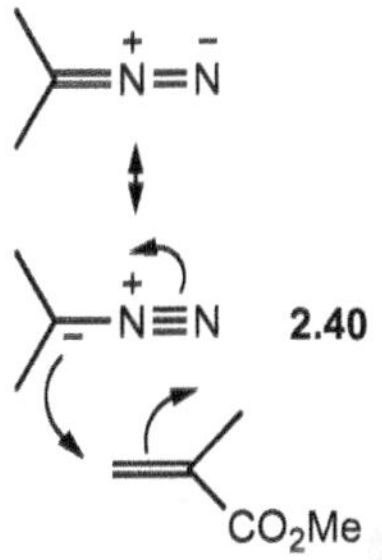

Arrows, therefore, do not reliably identify the nucleophilic and the electrophilic end of a dipole.

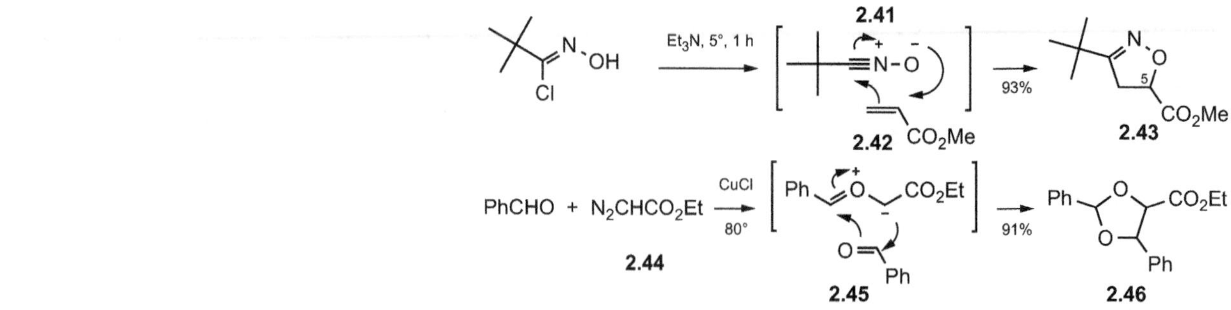

Fig. 2.5 The 1,3-dipolar cycloadditions of a nitrile oxide and a carbonyl ylid

dipolarophiles, but not with electron-poor dipolarophiles, and other dipoles are the other way round.

To make matters even more complex, the presence of substituents on the dipole can change these patterns, affecting the relative reactivity towards electron-rich and electron-poor dipolarophiles, and influencing the regioselectivity. Thus, the carbonyl ylid reaction **2.45** has a well-defined regiochemistry determined only by the substituents, since the core dipole is symmetrical. This reaction also illustrates the point that dipolarophiles do not have to be alkenes or alkynes—they can also have heteroatoms.

2.4 [4 + 2] Cycloadditions of cations and anions

Diels–Alder reactions are sometimes classified as [4 + 2] cycloadditions, and 1,3-dipolar cycloadditions as [3 + 2], where the numbers identify the number of *atoms* involved in the two chains. This classification is not as useful as the one used throughout this book, which is based on the number of *electrons* involved. In this classification, both Diels–Alder reactions and 1,3-dipolar cycloadditions are [4 + 2] cycloadditions—four electrons from the diene or dipole, and two from the dienophile or dipolarophile. These are not the only [4 + 2] cycloadditions, but they are by far the most numerous and the most important. Conjugated ions like allyl cations, allyl anions, and pentadienyl cations are all capable of cycloadditions. Thus, an allyl cation can be a two-electron component in a [4 + 2] cycloaddition, as in the reaction of the methallyl cation **2.48** derived from its iodide **2.47**, with cyclopentadiene giving a seven-membered-ring cation **2.49**. The diene is the four-electron component. The product eventually isolated is the alkene **2.50**, as the result of the loss of the neighbouring proton, the usual fate of a tertiary cation.

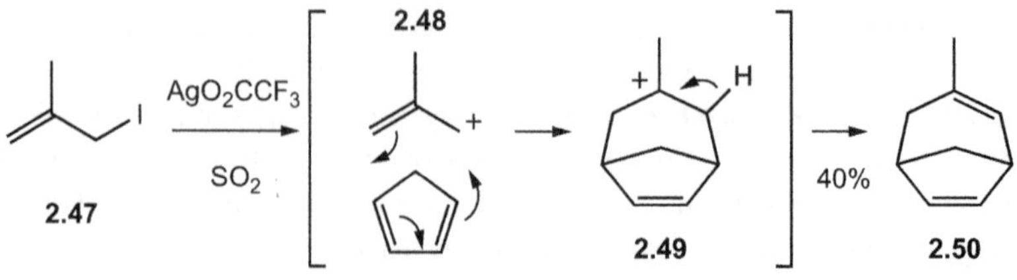

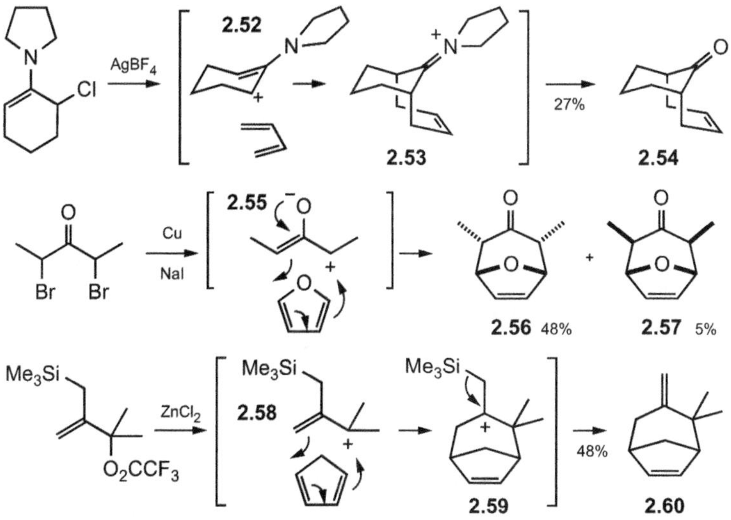

Fig. 2.6 Some cycloadditions of allyl cations

If both σ-bonds form at the same time, the reaction is pericyclic, but carbocations are capable of reacting with alkenes, and so the first bond may form to give the cyclopentenyl cation **2.51** in one step, with the second bond formed in a separate step **2.51** (arrows). The reaction remains a cycloaddition whether both bonds are formed at the same time or not, but it is pericyclic only if they are both formed in the same step, as is probable in this case. Fig. 2.6 shows a small selection of other cycloadditions involving allyl cations that may similarly be pericyclic.

If the reaction is pericyclic, the conjugated system of the allyl cation contracts to the unconjugated empty p-orbital of a simple carbocation; so this type of reaction is apt to be favourable only when the cationic centre in the product carries a substituent to stabilize it—the nitrogen lone pair in **2.52**, the oxygen lone pair in **2.55**, or the β-silyl group in **2.58** stabilize the cationic centres in the products **2.53**, **2.56** and **2.57**, and **2.59**, respectively. As in Diels–Alder reactions, dienes must be in the s-*cis* conformation for pericyclic cycloaddition to take place, but allyl cations, unlike most dienophiles, are reactive intermediates, rarely surviving long enough to meet a diene in the s-*cis* conformation. Open-chain dienes are therefore apt to give low yields, as in the formation of the bicyclic ketone **2.54**, whereas cyclic dienes are better, as in the formation of the tricyclic ketone **2.61**.

An allyl anion has two more electrons, and so it can participate in a cycloaddition as the four-electron component, which can react with an alkene in a [4 + 2] cycloaddition. The complication with anions is that they are often ill-defined, usually being organometallic species with a carbon–metal bond of substantial covalent character. It is not usually legitimate to picture conjugated anions, let alone draw them, simply as symmetrical conjugated systems of p-orbitals. Nevertheless, there does appear to be an element of concertedness in the reaction of

The cycloaddition of an allyl cation to a diene may not always be pericyclic.

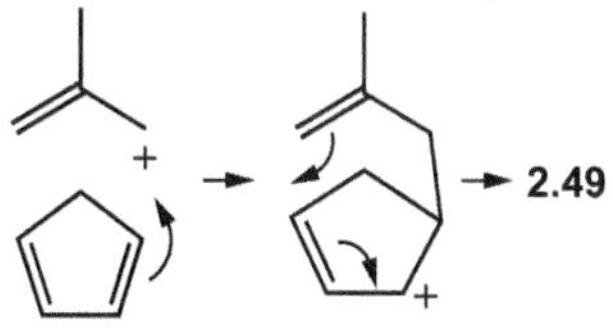

Cyclopentadiene, constrained to be s-*cis*, gives a higher yield of the tricyclic ketone **2.61** than butadiene gives of the bicyclic ketone **2.54**:

Some structures for 'allyl anions':

The reaction **2.66** could have been drawn:

and would still mean the same thing. The cationic carbonyl group in **2.66** is a highly stabilized cation, and the other methoxy group is a peripheral substituent, stabilizing the cationic centre in the product **2.67**. The three arrows shown above are the core reaction; the two extra arrows on **2.66** are substituent effects and do not count in assessing this reaction as a six-electron pericyclic process.

an allyl anion-like species, drawn as the 2-phenylallyl anion **2.63**, and prepared in an unfavourable equilibrium by treating α-methylstyrene with base. This anion undergoes a cycloaddition to an alkene such as stilbene **2.62**, present *in situ*, to give the cyclopentyl anion **2.64**, and hence the cyclopentane **2.65** after protonation.

A stepwise pathway for this reaction is less plausible than it was for the corresponding cationic reactions on p. 15—carbanions, whatever their actual nature, do not easily add to simple alkenes. The pericyclic pathway has the energetic benefit of forming both new σ-bonds in the same step, and so this type of reaction is quite plausibly pericyclic. Again, the anionic product **2.64**, having lost the allyl conjugated system, needs an anion-stabilizing group to make the reaction more favourable. This type of reaction is rare, maybe because allyl anions are not usually simple conjugated systems of p-orbitals; this makes it difficult for the overlap to develop at both ends simultaneously. It would be a powerful method for synthesizing cyclopentanes, if only it worked more generally.

Another rare kind of six-electron ionic cycloaddition is that between a pentadienyl cation and an alkene. A telling example is the key step **2.66** → **2.67** in a synthesis of gymnomitrol **2.68**, where the nature of the pericyclic step is heavily disguised, but all the more remarkable for that. Ionization of the acetal gives the cationic quinone system **2.66**. That this is a pentadienyl cation can be seen in the drawing of a canonical structure on the left, with the components of the pericyclic cycloaddition emphasized in bold. Intermolecular [4 + 2] cycloaddition takes place, with the pentadienyl cation as the four-electron component and the cyclopentene as the two-electron component. This reaction is an excellent example of how a reaction can become embedded in so much framework that its pericyclic nature is obscured.

2.5 Cycloadditions involving more than six electrons

All the reactions described so far have mobilized six electrons in the transition structure. Other numbers are possible, notably a few [8 + 2] and [6 + 4] cyclo-additions involving ten electrons in the cyclic transition structure. It is no accident, as we shall see in the next chapter, that these reactions have the same number of electrons (4n + 2) as aromatic rings.

A conjugated system of eight electrons would normally have the two ends of the conjugated system far apart, but there are a few molecules in which the two ends are held, more or less rigidly, close enough to participate in cyclo-additions to a double or triple bond. Thus, the tetraene **2.69** has the two ends of the conjugated system pulled together by the methylene bridge, allowing it to react with dimethyl azodicarboxylate **2.70** to give the [8 + 2] adduct **2.71**. Similarly, heptafulvalene **2.72** reacts with dimethyl acetylenedicarboxylate **2.73** to give the [8 + 2] adduct **2.74**, which is unstable, but loses hydrogen with the help of a palladium catalyst *in situ*, to give the relatively stable and easily isolated azulene **2.75**.

An [8 + 2] cycloaddition:

[6 + 4] Cycloadditions are a little more common than [8 + 2] cycloadditions, since it is a little easier to find pairs of conjugated systems that have the ends a suitable distance apart. Thus, in one of the first examples of such a reaction to be found, tropone **2.76** adds as a six-electron component to cyclopentadiene giving the adduct **2.77**. N-Ethoxycarbonylazepine **2.78** dimerizes, slowly at room temperature but rapidly on heating, to give mainly the adduct **2.79**, with one molecule acting as a six- and the other as a four-electron component.

A [6 + 4] cycloaddition:

Ordinary curly arrows, with their heads on an imaginary line joining the two atoms forming a new bond, would not be clear in **2.78**. The long arrows solve this problem, with the sacrifice of elegance, by passing through one of the atoms C-2 and C-5′, and pointing directly at the other C-2′ and C-7, identifying unambiguously that the new bonds are between C-2 and C-2′, and between C-5′ and C-7.

2.6 Allowed and forbidden reactions

Some forbidden reactions:

So far, we have seen [4 + 2], [8 + 2], and [6 + 4] cycloadditions, but they are not the only possibilities. Thus, butadiene dimerizes in a Diels–Alder reaction, but there are equally plausible-looking reactions, a [2 + 2] cycloaddition and a [4 + 4] cyclo-addition, which it could have undergone and did not. One might simply answer, in this case, that forming a six-membered ring is always easier than forming a four- or eight-membered ring, for reasons of ring strain and because of the high probability of the two ends being close. But this argument does not explain why ethylene and maleic anhydride, or many other combinations of alkenes, do not give a cyclobutane when heated together.

This is a deeply important point, and it is just as well that it is true—if alkenes and other double-bonded compounds could readily dimerize to form four-membered rings, there would be few stable organic molecules, and life would be impossible. It is not that the dimerization is energetically unprofitable—the four-membered ring is lower in energy than the two alkenes—so there must be a high kinetic barrier to the cycloaddition of one alkene to another. Chapter 3 explains where this barrier comes from.

A simple rule for thermal cycloadditions

Crudely, but adequately for now, we may state a rule governing which cycloadditions can take place and which cannot. A thermal pericyclic cyclo-addition is allowed if the total number of electrons involved can be expressed in the form $(4n + 2)$, where n is an integer. If the total number of electrons can be expressed in the form $4n$ it is forbidden. Another way of saying the same thing is that reactions with an odd number of curly arrows are allowed and those with an even number are forbidden. This rule needs to be qualified, as we shall see shortly, and in due course in Chapter 3 made more precise, along with the rules for all the other kinds of pericyclic reaction, in one all-encompassing rule. For now, we need to introduce the rule for photochemical pericyclic cycloadditions.

2.7 Photochemical cycloadditions

The simplified rule for photochemical cycloadditions is that they take place when the total number of electrons involved can be expressed in the form 4n, in other words when there is an even number of curly arrows. Thus, alkenes do give four-membered rings in photocycloadditions, either by self-coupling, or by cross-coupling. Ethylene and maleic anhydride, the very compounds that do not react on heating, give the cyclobutane **2.80** when irradiated with ultraviolet light. Likewise, dienes can undergo [4 + 4] cycloadditions, as in the intramolecular reaction **2.81** → **2.82**, and there are even a few [6 + 6] cycloadditions, as in the low-yielding photo-dimerization of tropone **2.76** to give the tricyclic adduct **2.83**.

A simple rule for cycloadditions in the first electronically excited state

A [2 + 2] photocycloaddition

A [4 + 4] photocycloaddition

A [6 + 6] photocycloaddition

However, photo-activation puts so much energy into the molecule, that many pathways become available to the first electronically excited state in addition to the relatively simple pericyclic change. For this reason, none of the photochemical reactions mentioned above can be guaranteed to be pericyclic, and all that one should take from these results is the very strong and suggestive contrast with the rules for thermal cycloadditions. This contrast is accentuated by the observation that photochemical Diels–Alder reactions are very rare, in spite of the ease with which six-membered rings are normally formed.

2.8 The stereochemistry of cycloadditions

In the course of a pericyclic cycloaddition, the p-orbitals at the ends of the conjugated system of each component form σ-bonds to the p-orbitals at the ends of the conjugated system of the other component. The p-orbitals must therefore approach each other head on, as in the transition structure for a

2.84

(a) Suprafacial bond formation   (b) Antarafacial bond formation

Fig. 2.7 Suprafacial and antarafacial defined for overlap developing to π-conjugated systems

Diels–Alder reaction **2.84** with the new σ-bonds forming between C-1 and C-1′ and between C-4 and C-2′. At the same time, a new π-bond forms between C-2 and C-3 of the diene. It is important to have this picture always in mind, since the usual flat drawings with curly arrows, like the one in Chapter 1, p. 2, disguise the geometry of approach. The two new bonds forming at C-1 and C-4 of the diene are both forming on the bottom surface of the conjugated system. Reactions in which the new bonds are being formed on the same surface are described as being **suprafacial** on that component (Fig. 2.7a). The two new bonds to C-1′ and C-2′ of the dienophile are also forming to the same surface of that π-bond, so this is also suprafacial. The opposite of suprafacial is called **antarafacial**, in which attack takes place with one bond forming to one surface but the other bond forming to the other surface (Fig. 2.7b). It is rare; it does not occur in any of the reactions illustrated here—nor shall we see antarafacial events for some pages yet.

The preliminary rules, thermal and photochemical, given in Section 2.6 on p. 18, need now to be qualified—they apply only to cycloadditions that are *suprafacial on both components*. Nevertheless, almost all pericyclic cycloadditions are suprafacial on both components. It is physically difficult for one conjugated system to suffer antarafacial attack from another, since it implies that one or another of the components can reach round from one surface to the other **2.85**. Only if at least one of the components has a long conjugated system can it twist enough to make this even remotely reasonable. Straightforward antarafacial attack in cycloadditions is therefore very rare indeed. Keep in mind, however, that these rules only apply to pericyclic cycloadditions—there are other kinds of cycloaddition, in which the two bonds are formed one at a time, to which none of these rules applies.

Since both the diene and the dienophile in Diels–Alder reactions **2.84** react suprafacially, there are synthetically useful structural consequences, which incidentally prove that the stereochemistry of approach is suprafacial. Thus, a diene with substituents at each end forming new bonds to one surface will create an adduct in which the stereochemical relationships are preserved. The *trans,trans*-diene **2.86** and diethyl acetylenedicarboxylate **2.87** give the adduct **2.88**, in which the two phenyl substituents are *cis* on the cyclohexadiene ring. Visualizing this relationship takes a little practice, but if the product, **2.88** drawn to resemble the transition structure, is looked at from the right-hand side, the two hydrogen atoms are both coming towards the viewer, and the two phenyl groups are behind, as redrawn in **2.89**. We could equally easily have drawn the dienophile attacking from the top surface, when the product would have the two phenyl

Antarafacial overlap on one component in a cycloaddition would need most unusually long and flexible conjugated systems:

antarafacial component
suprafacial component

2.85

groups coming towards the viewer. This is simply the enantiomer of **2.89**. Since the reaction is between two achiral components, it must give an equal amount of each. The important point is that the suprafacial development of overlap to the diene **2.86** produces the product with the two phenyl groups *cis* to each other.

A suprafacial reaction on a diene

Likewise with the dienophile: the maleate and fumarate esters **2.90** and **2.92** react with butadiene to give diastereoisomeric adducts **2.91** and **2.93**, in which the substituents retain, as a consequence of the suprafacial nature of the developing overlap, the *cis* and *trans* relationships they had in the dienophiles. Diels–Alder reactions are much used in organic synthesis, not only because two new C–C bonds are made in one step, but also because the relative stereochemistry of up to four new stereogenic centres may be controlled in a highly predictable sense.

Reactions like the cycloadditions **2.90**→**2.91** and **2.92**→**2.93**, where one compound gives one product, but its stereoisomer gives a stereoisomeric product, are described as being **stereospecific**. Notice that a reaction can only be proved to be stereospecific when both isomers of the starting material are tested in order to be sure that they give stereochemically different products.

All the other kinds of pericyclic cycloaddition discussed so far, not just Diels–Alder reactions, are also suprafacial on both components. Thus, 1,3-dipolar cycloadditions involve suprafacial attack on the dipolarophile, as in the reaction of diazomethane on the dimethylated fumarate **2.94** and maleate **2.96** esters, which give stereospecifically the *trans* and *cis* adducts **2.95** and **2.97**, respectively. In contrast, it is impossible to demonstrate the suprafacial nature of the attack on most open-chain dipoles—diazoalkanes do not and cannot develop identifiable stereochemistry at the nitrogen end of the dipole. Azomethine ylids, however, do have the possibility of demonstrable stereochemistry at both ends, and they do react stereospecifically and suprafacially, as in the reaction of the dipoles **2.98** and **2.100**, which give the diastereoisomeric adducts **2.99** and **2.101** with dimethyl acetylenedicarboxylate. The carbonyl ylids **2.98** and **2.100** are reactive intermediates prepared in stereospecific electrocyclic reactions (Chapter 4, p. 74).

The π orbitals of the ylids **2.98** and **2.100** are isoelectronic with those of an allyl anion, in which ψ_1 produces π-bonding on both sides of the central atom, restricting rotation.

2.94   **2.95**   **2.96**   **2.97**

2.98   **2.99**   **2.100**   **2.101**

All the other cycloadditions, such as the [4 + 2] cycloadditions of allyl cations and anions, and the [8 + 2] and [6 + 4] cycloadditions of longer conjugated systems, have also been found to be suprafacial on both components, wherever it has been possible to test them. Thus the *trans* phenyl groups on the cyclopentene **2.65** show that the two new bonds were formed suprafacially on the *trans*-stilbene. The tricyclic adducts **2.61**, **2.77**, **2.79**, and **2.83**, and the tetracyclic adduct **2.82**, show that both components in each case have reacted suprafacially, although only suprafacial reactions are possible in cases like these, since the products from antarafacial attack on either component would have been prohibitively strained. Nevertheless, the fact that they have undergone cycloaddition is important, for it is the failure of thermal [2 + 2], [4 + 4], and [6 + 6], and photochemical [4 + 2], [8 + 2], and [6 + 4] pericyclic cycloadditions to take place that is significant, even when all-suprafacial options are open to them.

'Extended'
transition
structure

'Compressed'
transition
structure

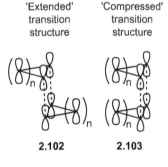

2.102   **2.103**

Compressed transition structures are usually preferred in allyl cation + diene reactions

There are other stereochemical features that have nothing to do with the symmetry of the orbitals, and are much less powerfully controlled. In many cycloadditions, there are two possible all-suprafacial approaches: one having what is called the extended transition structure **2.102**, in which the conjugated systems keep well apart, and the other called the compressed **2.103**, where they lie one above the other. Both are equally allowed by the rules that we shall see in Chapter 3, but one will usually be faster than the other. This type of stereochemistry applies only when the conjugated systems have at least three atoms in each component; it is therefore only rarely a consideration. It shows up in the cycloadditions of allyl cations to dienes, where the two adducts **2.56** and **2.57** in Section 2.4 on p. 15 are the result of the compressed transition structure **2.104** and the extended **2.105**, respectively, with the former evidently lower in energy.

major   minor

2.104   **2.56**   **2.105**   **2.57**

In contrast, the tricyclic ketone **2.77** and the azepine dimer **2.79** from Section 2.5, pp. 17 and 18 are the products of attack with extended transition structures **2.106** and **2.107**; in neither case is the alternative product detected. Clearly, different systems use different rules. In the cycloadditions of these longer conjugated systems, the influence of the substituents on the conjugated system must also be playing some part, and might be overriding the inherent preferences for extended or compressed transition structures. This area is not well enough understood for anyone to be dogmatic about what the underlying preferences ought to be in these comparatively rare cycloadditions.

$$\text{2.106} \longrightarrow \text{2.77}$$

$$\text{2.107} \longrightarrow \text{2.79}$$

Extended transition structures are usually preferred in [6 + 4] cycloadditions.

The effect of substituents is usually more important, because it can apply to all conjugated systems, not just those with more than two atoms. When a conjugated system carries an electron-withdrawing substituent Z on one side, as it often does in Diels–Alder reactions, that substituent may sit pointing away from the conjugated system of the other component **2.108**, or it may sit under it **2.109**. The former, the extended transition structure, is called the *exo* mode of attack, and the latter, the compressed transition structure, is called *endo*. The *endo* approach suffers from a steric repulsion between the group Z and the orbitals of the conjugated system. For this reason, the *exo* approach can be expected to lead to the major isomer, and in many cases it does.

However, Diels–Alder reactions are well known to be exceptional, with maleic anhydride reacting with cyclopentadiene by way of an *endo* transition structure **2.110** to give what is called the *endo* adduct **2.111** as the major product. The *exo* adduct **2.112** is a very minor product, unless the mixture is heated for a long time, when reversal of the Diels–Alder reaction and re-addition establish the thermodynamic equilibrium in its favour. The *endo* adduct is evidently the product of kinetic control, and the preference for it, established by Alder, is called the **endo rule**.

Exo attack:

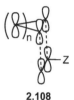

2.108

Endo attack:

2.109

$$\text{2.110} \xrightarrow{\text{r.t.}} \text{2.111} \underset{}{\overset{190°}{\rightleftharpoons}} \text{2.112}$$

Alder's *endo* rule applies not only to cyclic dienes like cyclopentadiene and to disubstituted dienophiles like maleic anhydride, but also to open chain dienes and to mono-substituted dienophiles: diphenylbutadiene and acrylic acid, for example, react by way of an *endo* transition structure **2.113** to give largely (9:1) the adduct **2.114** with all the substituents on the cyclohexene ring *cis*, and equilibration again leads to the minor isomer **2.115** with the carboxyl group *trans* to the two phenyl groups.

Alder's *endo* rule leads substituents in open-chain *trans* dienes to be all-*cis* on the cyclohexene ring in the kinetically favoured product.

2.113   **2.114**   **2.115**

The *endo* rule also applies to some Diels–Alder reactions with inverse electron demand, as in the cycloaddition of butadienylsulfoxide **2.116** with the enamine **2.117**, which gives only the adduct **2.118**. The amino group is an electron-donating substituent on the dienophile, and the sulfoxide is an electron-withdrawing group on the diene, and they appear *cis* to each other in the product consistent with an *endo* transition structure.

A Diels–Alder reaction with inverse electron demand:

2.116   **2.117**   **2.118**

1,3-Dipolar cycloadditions are not so straightforward. More often than not the selectivity is low, as in the cycloadditions **2.119** and **2.121** of methyl acrylate to *C,N*-diphenylnitrone, which give the adducts **2.120** and **2.122** only marginally in favour (57:43) of the *exo* mode. However, the *endo* mode is sometimes favoured in other 1,3-dipolar cycloadditions in ways that might be synthetically useful, and sometimes it is the *exo*, depending in a not always predictable way upon the dipole, the dipolarophile and their substituents.

Dipolar cycloadditions often show low levels of *endo*, *exo* stereoselectivity:

2.119   **2.120**   **2.121**   **2.122**

For example, the cycloaddition **2.123** of an azomethine ylid to dimethyl maleate gives largely (3:1) the *endo* adduct **2.124**, whereas the cycloaddition **2.126** of a nitrone to a vinyl ether gives largely (92:3) the *exo* adduct **2.127**. In experiments like these, the nature of the substituents changes, and one is not comparing like with like. It is not even clear in every case what the geometry of the dipole is, or whether the results are kinetically or thermodynamically controlled. In contrast to Diels–Alder reactions, where some general principles obtain, it is advisable to look up close analogies before using the stereoselectivity of a 1,3-dipolar cyclo-addition in synthetic planning.

Endo attack is preferred in some 1,3-dipolar cycloadditions.

Exo attack is preferred in some others.

2.9 Regioselectivity of cycloadditions

In addition to the stereoisomers, already discussed, unsymmetrical dienes like piperylene **2.129** or isoprene **2.133** and unsymmetrical dienophiles like methyl acrylate **2.130** give two adducts, depending upon the orientation of the two substituents in the product (see p. 10). The selectivity for the formation of one adduct, say **2.131**, over the other **2.132** is called **regioselectivity**, and the iso-mers are called regioisomers. The degree of regioselectivity is often high, even with substituents like methyl that are not strongly polar, and especially when the lowest possible temperature is used, as in the examples below.

The major adducts from mono-substituted dienes and dienophiles are somewhat loosely but understandably referred to as 'ortho' and 'para'.

In general, 1-substituted dienes and mono-substituted dienophiles react to give more of the 'ortho' adduct with the substituents on adjacent atoms, and 2-substituted dienes almost always react to give more of the 'para' adduct having the substituents on diametrically opposite atoms. The extraordinary feature of this pattern is that it is true for almost every combination of electron-donating (X-) and electron-withdrawing substituent (Z-), or even simply from extended conjugation like vinyl and phenyl (C-). The only exception is the almost unknown combination of an X-substituted diene and an X-substituted dienophile, which is marginally 'meta' selective.

In order to explain this pattern, it is common to oversimplify with an example like the addition of methoxybutadiene **2.138** to acrolein **2.139**, giving only the 'ortho' adduct **2.140**, and to identify C-4 of the diene as having partial negative charge (**2.136**, arrows), which will attract the partial positive charge on the β carbon of the acrolein (**2.137**, arrows). Generally, the more powerful the electron-donating and electron-withdrawing substituents, the more regioselective is the reaction. This explanation also works for the example given previously of inverse electron demand, with C-4 of the diene **2.116** electrophilic and the β carbon of the enamine system **2.117** nucleophilic.

2.136

2.137

2.138   **2.139**   100°, 3 h   60%   **2.140**

Although this idea is not without merit, it is not the whole explanation—butadiene-1-carboxylic acid **2.142** and acrylic acid **2.143** also give mainly the 'ortho' adduct **2.144**. Since butadiene-1-carboxylic acid can be expected to have a partial positive charge on C-4 (**2.141**, arrows), that atom ought to be repelled by the partial positive charge on the β carbon of the acrylic acid. A better explanation for the regioselectivity will be given on pp. 59 and 60 in Chapter 3.

2.141

2.142   **2.143**   150°   86%   **2.144**   90:10

Regioselectivity in 1,3-dipolar cycloadditions is more complicated. We saw in Figs. 2.4 and 2.5 in Section 2.3 on pp. 13 and 14, some examples of unsymmetrical dipoles and unsymmetrically substituted dipoles showing selectivity in the orientation of adduct formation with mono-substituted dipolarophiles. These patterns were already not as amenable to simple explanation as Diels–Alder reactions, since it is not immediately obvious which end of a dipole is nucleophilic and which electrophilic—the way the formal charges are drawn is no guide, as we saw with diazopropane **2.40**. We also saw that some dipoles like phenyl azide are consistent, reacting to give opposite regioisomers **2.33** and **2.35** with dipolarophiles that have electron-withdrawing and electron-donating substituents.

Other dipoles, however, are inconsistent, with a nitrone reacting with both kinds of dipolarophile with the same regiochemistry placing the substituent on C-5 in both adducts **2.37** and **2.39**. Even that pattern does not hold up when the electron-withdrawing power is increased—N-methyl-C-phenylnitrone reacts with diethyl methylenemalonate to give the adduct **2.145**, the opposite regio-isomer to the adduct **2.36** it gives with methyl acrylate.

With so many variables to contend with—large numbers of dipoles and dipolarophiles, seemingly unpredictable and not always consistent patterns of reactivity and regiochemistry, and substituent effects that can support or override the core patterns—it is not surprising that almost no one can master 1,3-dipolar cycloaddition chemistry. For the present it is enough for you to recognize a 1,3-dipolar cycloaddition when you see it, and to appreciate the nature of some of the problems that exist.

2.145

2.10 **Intramolecular cycloadditions**

Intramolecular reactions are obliged to obey the powerful rules that govern which cycloadditions are allowed and which are forbidden, but they can over-ride the weaker forces controlling regio- and stereoselectivity, and it is important to remember this opportunity for synthetic design, when the trends outlined previously are in conflict with the aims of the synthesis.

For example, the electrocyclic opening of the benzcyclobutene **2.146** gives a diene unit, which is trapped intramolecularly by the pendant vinyl group in an *exo* transition structure **2.147**. An intermolecular version of this reaction would, in the first place, be too slow, because of the absence of activating substituents. In the second place, it would have an *endo* transition structure. Finally, there would be little regiocontrol, since the electron-donating power of the substituents is feeble, and, if anything, probably in favour of the regioisomer. The presence of the chair-like con-necting chain forces the stereochemistry and the regiochemistry into that required for a beautifully simple steroid synthesis at a high but now practical temperature.

2.146   **2.147**   **2.148**

Similarly, the nitrone dipole **2.151** spontaneously adds obediently and supra-facially to the alkene, but with regiocontrol stemming from the intramolecular nature of the reaction, rather than from any substituent effects, since the double bond is essentially symmetrical. Bond formation for 1,3-dipolar cycloadditions is inherently asynchronous, with the C–C bond in this case forming ahead of the

2.149   **2.150**

O–C bond (asynchronicity is discussed again on p. 59 in Chapter 3). Since the C–C bond is the more strongly developed, it is easier if it starts to form a six-membered ring **2.149** than if it starts to form a seven-membered ring **2.150**. Again the intramolecularity completely controls the regiochemistry, and contributes to the success of a short synthesis of luciduline **2.153**.

2.151   51%   **2.152**   **2.153**

2.11 Not all cycloadditions are pericyclic

A cation-stabilizing group on one component and an anion-stabilizing group on the other can make a stepwise reaction feasible:

We have already seen in the reactions of an allyl cation with a diene, that cycloadditions may take place stepwise, with one bond forming in a separate step from the second, in which case, although they are still cycloadditions, they are no longer pericyclic. It is not absolutely certain that all the cycloadditions illustrated previously are pericyclic. All that is required for a stepwise reaction to be plausible is that the intermediate produced when one bond forms ahead of the other should have substituents powerful enough to stabilize the charges.

   Most of the examples in this chapter so far probably do not have this feature, with the possible exception of the Diels–Alder reaction with inverse electron demand **2.117** → **2.118**, and are therefore likely to be pericyclic cycloadditions. But it is not difficult to set up a system in which one bond can form before the other—it is merely necessary to equip one component with a powerful electron-donating group to make it nucleophilic, and the other with a powerful electron-withdrawing group to make it electrophilic. Thus, the enamine **2.154** readily reacts with methyl vinyl ketone **2.155** to give the hetero Diels–Alder adduct **2.157**. However, this reaction is not pericyclic, and there is ample evidence for an intermediate **2.156** produced only by the nucleophilic attack of C-2 of the enamine on the β-carbon of the enone system. Enamines are well known to be nucleophilic enough to react with electrophilic alkenes, and do not need to make a second bond at the same time for reaction to be energetically feasible. Indeed, in the corresponding reaction with methyl acrylate **2.158**, a six-membered ring is not formed, and the product is a cyclobutane **2.160** in a two-step process by way of the zwitterionic intermediate **2.159** analogous to the intermediate **2.156**. There are a lot of reactions of this type—they are cycloadditions but not pericyclic, and the rules do not apply to them.

A stepwise ionic reaction looking like a Diels–Alder reaction:

2.154   **2.155**   r.t., 20 min   **2.156**   60%   **2.157**

2.154   2.158   2.159   2.160

A stepwise ionic reaction looking like a forbidden [2 + 2] pericyclic cycloaddition

Stepwise reactions by way of diradical intermediates are also possible; they often require rather high temperatures, but they are probably involved in the formation of cyclobutanes by the thermal coupling of alkenes like the halogenated alkene **2.161** with themselves or with dienes like butadiene giving the cyclobutane **2.163**. The radical centres in the intermediate **2.162** are stabilized, the one on the left by the α-chlorines and the β-fluorines, and the one on the right because it is allylic. There are a number of reactions like this—all that is required is enough radical-stabilizing substituents.

Radical-stabilizing groups can also make a stepwise reaction feasible:

2.161   **2.162**   **2.163**

An interesting question is, why isn't a six-membered ring produced from the intermediate diradical **2.162**? The answer is related to something mentioned earlier—the need for the diene to be in the relatively high energy s-*cis* conformation for the Diels–Alder reaction. For the first step in a stepwise pathway, the diene does not need to be in the s-*cis* conformation—as long as the substituents stabilize the radicals well enough, as they do here, the first bond can form the diradical **2.162**, redrawn here as **2.164**, while the diene is still in its s-*trans* conformation. The allyl radical **2.164** produced from the s-*trans* diene has itself a fixed configuration, in which rotation about the bond between C-2 and C-3 is more restricted than it was in the diene. The chlorine-stabilized radical does not attack this radical at C-1 (**2.164**, arrows), where there is an equal probability of finding the unpaired electron, because, were it to do so, it would have to give a *trans*-cyclohexene **2.165**. Rotation about the bond between C-2 and C-3 is evidently too slow to compete with the radical recombination step **2.162** → **2.163**.

There is a substantial fraction of a π-bond between C-2 and C-3, as well as between C-1 and C-2, in an allyl radical, just as there was in the allyl anion in Section 2.8 on p. 21:

2.164   **2.165**

Since stepwise reactions are not subject to the rules of pericyclic reactions, they are often invoked to explain how reactions in which the rules have been subverted take place. However, there is a small group of thermal [2 + 2] cycloadditions that seem to be disobeying the rules, and yet may well be pericyclic. One is the reaction of ketenes with electron-rich alkenes, illustrated by the reaction of diphenylketene **2.167** with ethyl vinyl ether **2.166** giving the cyclobutanone **2.168**. Another is a group of electrophile-induced dimerizations of allenes and acetylenes, all involving a vinyl cation intermediate that is undergoing cycloaddition to an alkene or alkyne, such as the Smirnov–Zamkov reaction of dimethylacetylene **2.169** with chlorine, which gives the vinyl cation **2.170**, and hence the dichlorocyclobutene **2.172**. These [2 + 2] cycloadditions may well be pericyclic, but we shall see in Chapter 3 that the question of why they occur needs special treatment.

Some anomalous reactions looking like forbidden [2 + 2] pericyclic cycloadditions

They are both reactions of vinyl cations, either as such **2.170**, or highly stabilized vinyl cations in the case of ketene **2.167**, where the carbonyl group is a stabilized carbocation.

2.12 Cheletropic reactions

A number of cheletropic reactions also appear to be anomalous, including the best known of all cheletropic reactions, the stereospecific insertion of a carbene into a double bond, as in the reaction of dichlorocarbene **2.173** with alkenes. Here, we have a reaction involving only four electrons, which is known to be suprafacial on the alkene, preserving the geometry of the substituents in the starting alkenes in the cyclopropanes **2.174** and **2.175**. Furthermore, the [2 + 2] reaction takes place even with a diene, which could undergo an allowed [4 + 2] reaction, but chooses not to.

An anomalous pair of cheletropic reactions looking like forbidden [2 + 2] suprafacial cycloadditions:

We shall find that this reaction is again a rather special anomaly, needing special treatment, but there are straightforward six-electron cheletropic reactions, such as the irreversible extrusion of nitrogen from the diazene **2.176**, and the easy loss of carbon monoxide from norbornadienone **2.177**.

2.176   **2.177**

A pair of allowed suprafacial retro-[4 + 2]-cycloadditions

Reactions like these, where they have been tested, have also proved to be suprafacial on the conjugated system. Thus, the reversible insertions of sulfur dioxide into the stereochemically labelled butadienes **2.178** and **2.179** take place stereospecifically, and suprafacially.

Stereospecific cheletropic reactions

2.178   **2.179**

There are also some eight-electron cheletropic reactions that follow an allowed pathway. The most striking are retro-cycloadditions, but they have an intriguing feature. In the stereospecific extrusion of sulfur dioxide from the seven-membered ring sulfones **2.181** and **2.183** we see for the first time a reaction in which one of the components is acting in an antarafacial manner. This can be appreciated most easily by looking at the reverse reaction. Imagine the cycloaddition of sulfur dioxide to the *trans,cis,trans*-triene **2.180**, where the stereochemistry giving the *trans* isomer **2.181** must involve overlap developing **2.184**, one bond to the top side of the conjugated triene and one to the bottom—an antarafacial step. The reaction takes place in the opposite direction to this, but the pathway is the same.

An antarafacial reaction to explain its reverse:

2.180   **2.181**   **2.182**   **2.183**   **2.184**

2.13 **Further reading**

The following books and articles, and the chapters in *Comprehensive Organic Synthesis*, Vol. 5, ed. L. A. Paquette, Pergamon, Oxford, 1991 (referred to as *COS*), have detailed information:

W. Carruthers, *Cycloaddition Reactions in Organic Synthesis*, Pergamon, Oxford, 1990.

F. Fringuelli and A. Tatticchi, *Dienes in Diels–Alder Reactions*, Wiley, New York, 1990.

Advances in Cycloaddition, ed. D. P. Curran, JAI Press, Greenwich CT, Vol. 1, 1988, Vol. 2, 1990, and Vol. 3, 1993.

Intermolecular Diels–Alder:

W. Oppolzer in *COS*, Ch. 4.1; J. Hamer, *1,4-Cycloaddition Reactions*, Academic Press, New York, 1967.

D. L. Boger and S. M. Weinreb, *Hetero Diels–Alder Methodology in Organic Synthesis*, Academic Press, New York, 1967.

B. Weinreb in *COS*, Ch. 4.2; D. L. Boger in *COS*, Ch. 4.3.

Intramolecular:

D. F. Taber, *Intramolecular Diels–Alder and Alder Ene Reactions*, Springer, New York, 1984.

W. R. Roush in *COS*, Ch. 4.4.

retro:

R. W. Sweger and A. W. Czarnik in *COS*, Ch. 4.5.

B. Rickborn in *Org. React.* (NY), 1998, **52**, 1.

Inverse electron demand:

J. Sauer and H. Wiest, *Angew. Chem., Int. Ed. Engl.*, 1962, **1**, 268.

D. L. Boger and M. Patel, in *Progress in Heterocyclic Chemistry*, ed. H. Suschitzky and E. F. V. Scriven, Pergamon, Oxford, 1989, Vol. 1, p. 30.

Dipolar cycloadditions:

R. Huisgen, *Angew. Chem., Int. Ed. Engl.*, 1963, **2**, 565 and 633.

A. Padwa, ed. *1,3-Dipolar Cycloaddition Chemistry*, Wiley, New York, Vol. I and Vol. II, 1984.

A. Padwa in *COS*, Ch. 4.9.

[4 + 3] Cycloadditions:

J. H. Rigby and F. C. Pigge, *Org. React. (NY)*, 1997, **51**, 351.

[4 + 4] and [6 + 4] Cycloadditions:

J. H. Rigby in *COS*, Ch. 5.2.

[3 + 2] and [5 + 2] Photocycloadditions:

P. A. Wender, L. Siggel, and J. M. Nuss in *COS*, Ch. 5.3.

Thermal cyclobutane formation:

J. E. Baldwin in *COS*, Ch. 2.1, and Ch. 5 in: *Pericyclic Reactions*, ed. A. P. Marchand and R. E. Lehr, Vol. 2, Academic Press, New York, 1977.

Photochemical cycloadditions:

M. T. Crimmins and T. L. Reinhold, *Org. React. (NY)*, 1993, **44**, 297.

Cheletropic:

W. L. Mock, Ch. 3 in: *Pericyclic Reactions*, loc. cit.

W. M. Jones and U. H. Brinker, Ch. 3 in: *Pericyclic Reactions*, ed. A. P. Marchand and R. E. Lehr, Vol. 1, Academic Press, New York, 1977.

2.14 **Problems**

2.1 Explain why having a substituent on C-2 of a diene increases its reactivity in Diels–Alder reactions, and why *trans*-piperylene **2.9** is more reactive in Diels–Alder reactions than *cis*-piperylene.

2.2 Classify the following cycloadditions, and say whether they obey the rule for thermal pericyclic reactions:

(a)

(b)

(c)

heat

2.3 Identify the steps in these heterocyclic syntheses:

2.4 What cycloadditions would you use to synthesize these two compounds?

2.5 Explain why the dimerization of cyclobutadiene is the fastest known bimolecular pericyclic reaction:

$$2 \times \quad \xrightarrow{60K}$$

2.15 **Summary**

- Cycloadditions are those reactions in which two molecules with π-conjugated systems couple across their ends to form a ring; the reactions are pericyclic if both new bonds are formed at the same time.

- The reverse reaction is called a cycloreversion or a retro-cycloaddition.

- If the conjugated systems are a diene and an alkene, the reaction is called a Diels–Alder reaction.

- Diels–Alder reactions take place faster when the diene has electron-donating substituents and the dienophile has electron-withdrawing substituents, with the latter more important.

- If one or more of the conjugated atoms in a Diels–Alder reaction is an electronegative heteroatom, like, N, O, or S, the reaction is called a hetero-Diels–Alder reaction.

- If one of the conjugated systems has three atoms taken from the common elements C, N, O, and S and four π-electrons, the reaction with an alkene, carbonyl group, or imine is called a dipolar cycloaddition.

- Cationic and anionic conjugated systems can take part in cycloadditions.

- Pericyclic cycloadditions are energetically favourable when the total number of π-electrons involved can be expressed in the form [4n + 2]; the most common are those with six π-electrons, but reactions with two or ten π-electrons are well-known.

- Most thermal cycloadditions in which the total number of π-electrons can be expressed in the form [4n] are stepwise, either stepwise ionic if anion- and cation-stabilizing groups are present, or stepwise radical if radical-stabilizing groups are present.

- Photochemical cycloadditions are energetically favourable when the total number of π-electrons involved can be expressed in the form [4n]; the most common are [2 + 2]-cycloaddition with a total of four π-electrons, but reactions with eight π-electrons are well known.

- Unsymmetrical conjugated systems lead to products in which the two components are oriented in two different ways, called regioisomers.

- There is a high level of stereospecificity, stemming from the all-suprafacial attack on both components in almost all pericyclic cycloadditions.

- There is sometimes, especially in Diels–Alder reactions, a high level of stereoselectivity in favour of the *endo* adduct.

- Ketenes and carbenes undergo [2 + 2] cycloadditions that appear to be anomalous, but have the features of a pericyclic reaction.

3
The Woodward–Hoffmann rules and molecular orbitals

3.1 Introduction

The characteristic feature of all pericyclic reactions is the concertedness of all the bond making and bond breaking, and hence the absence of any intermediates. Naturally, organic chemists have worked hard, and devised many ingenious experiments to prove that pericyclic reactions are indeed concerted, concentrating especially on the Diels–Alder reaction.

Because pericyclic reactions are concerted there are powerful rules about which reactions are feasible and which not. The rules themselves, and their basis in the conservation of symmetry of the molecular orbitals, provide the material in the rest of this chapter.

3.2 Evidence for the concertedness of bond making and breaking

The Arrhenius parameters for Diels–Alder reactions show that there is an exceptionally high negative entropy of activation, typically in the range -150 to -200 J K^{-1} mol^{-1}, with a low enthalpy of activation reflecting the exothermic nature of the reaction. Bimolecular reactions inherently have high negative entropies of activation, but this is exceptionally high. The extra organization necessary for the two components to approach favourably aligned for *both* bonds to form at the same time accounts for the high value in pericyclic cycloadditions. The compact transition structure is also in agreement with the negative volumes of activation measured by carrying out the reaction under pressure.

Arrhenius parameters

The rates of Diels–Alder reactions are little affected by the polarity of the solvent. If a zwitterionic intermediate were involved, the intermediate would be more polar than either of the starting materials, and polar solvents would solvate it more thoroughly. Typically, a large change of solvent dipole moment, from 2.3 to 39, causes an increase in rate by a factor of only 10. In contrast, stepwise ionic cycloadditions take place with increases in rate of several orders of magnitude in

Small solvent effects

3.1

Additive or more than additive effects

Theoretical calculations

Rule-obeying pattern

polar solvents. This single piece of evidence rules out stepwise ionic pathways for most Diels–Alder reactions, and the only stepwise mechanism left is that involving a diradical.

Deuterium substitution on the four carbon atoms changing from trigonal to tetrahedral as the reaction proceeds, gives rise to inverse secondary kinetic isotope effects, small but measurable, both for the diene and the dienophile **3.1**. If both bonds are forming at the same time, the isotope effect when both ends are deuterated is geometrically related to the isotope effects at each end. If the bonds are being formed one at a time, the isotope effects are arithmetically related. It is a close call, but the experimental results, both for cycloadditions and for cycloreversions, suggest that they are concerted. This is known as the Thornton test.

Another way of testing how one end of the dienophile affects the other end is to load up the dienophile with up to four electron-withdrawing groups and see how each additional group affects the rate. Thus, a stepwise reaction for tetracyanoethylene ought not to take place much more than statistically faster than with 1,1-dicyanoethylene, but a concerted reaction ought to take place much faster. Furthermore, the relative rates can be compared with the rate of nucleophilic attack on the dienophile as a model for a stepwise reaction. Experiments based on this approach support a concerted mechanism.

High level molecular orbital calculations have been carried out, with ever increasing levels of sophistication. Most, but not quite all, suggest that the concerted pericyclic pathway gives the lowest energy transition structure. Calculations even more strongly support the evidence for concertedness from the isotope effect studies. A diradical intermediate would have an allylic radical at one end of the diene and the configuration would not have changed from trigonal to tetrahedral. The two isotope effects, at the end undergoing bonding and at the end carrying the odd electron, should therefore be very different. Since they are not different, but more than arithmetically reinforce each other, the reaction is most probably concerted.

Finally, the single most impressive piece of evidence comes from the very fact that pericyclic reactions obey the rules that we have seen already in the last chapter, and are about to expand upon. These rules only apply if the reactions are concerted. To have a few reactions accidentally obeying the rules would be reasonable, but to have a very large number of reactions seemingly falling over themselves to obey strict stereochemical rules, sometimes in the most improbable circumstances, is overwhelmingly strong evidence about the general picture. Of course, no single reaction can be proved to be pericyclic just because it obeys the rules—obedience to the rules is merely a necessary condition for a reaction to be accepted as pericyclic. In the case of cycloadditions, the suprafacial nature of the reaction on both components in a very high proportion of reactions at least says that the second bond, if it is not formed at the same time as the first, is formed very quickly after the first, before any rotations about single bonds can take place. It seems more than likely that most reactions thought to be pericyclic actually are.

We now turn to the ideas, based on molecular orbital (MO) theory, which have been advanced to explain the patterns of reactivity that all pericyclic reactions show.

3.3 The aromatic transition structure

The first and the most simple is the observation that the common thermally induced reactions have transition structures involving a total of $(4n + 2)$ electrons. We saw in the last chapter that $[4 + 2]$, $[8 + 2]$, and $[6 + 4]$ thermal cycloadditions are common, and that $[2 + 2]$, $[4 + 4]$, and $[6 + 6]$ cycloadditions are almost only found in photochemically induced reactions. The total numbers of electrons in the former group are all $(4n + 2)$ numbers, analogous to the number of electrons in aromatic rings.

A problem with this explanation is that it is a bit more difficult to explain those pericyclic reactions that we shall come to in Chapter 4, which smoothly take place in spite of their having a total of $4n$ electrons. We shall find that these all show stereochemistry involving an antarafacial component. It is possible to include this feature in the aromatic transition structure model—if the p-orbitals that make up a cyclic conjugated system have a single twist, like a Möbius strip, then the appropriate number of electrons for an aromatic system becomes $4n$ rather than $(4n + 2)$. An antarafacial component in a transition structure is equivalent to a **Möbius conjugated system** (see **2.85** in Chapter 2, on p. 20 for a hypothetical example, and **2.184** in Chapter 2 on p. 31 for a real example).

Although wonderfully simple, this explanation is not very revealing.

Aromatic transition structures are stabilized just as aromatic rings are

3.4 Frontier orbitals

The easiest explanation is based on the **frontier orbitals**—the highest occupied molecular orbital (HOMO) of one component and the lowest unoccupied orbital (LUMO) of the other. These orbitals are the closest in energy and make a disproportionate contribution to lowering the energy of the transition structure as they interact. Thus, if we compare a $[2 + 2]$ cycloaddition **3.2** with a $[4 + 2]$ cycloaddition **3.3** and **3.4**, we see that the former has frontier orbitals that do not match in sign at both ends, whereas the latter do, whichever way round, **3.3** or **3.4**, we take the frontier orbitals. The signs of the atomic wave functions

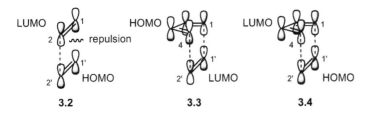

are given by the shading in the lobes of the p-orbitals throughout this book, to avoid any confusion with charge. In the [2 + 2] reaction **3.2**, the lobes on C-1 and C-1' are opposite in sign and represent a repulsion—an antibonding interaction. There is no barrier to formation of the bond between C-2 and C-2', making stepwise [2 + 2] cycloadditions like that between the enamine **2.154** and acrylic ester **2.158** in Chapter 2 on p. 29 possible; the barrier is only there for *both* bonds forming at the same time. The [4 + 4] and [6 + 6] cycloadditions have the same problem, and the [8 + 2] and [6 + 4] do not. Try it for yourself.

If we look again at the frontier orbitals for the [2 + 2] reaction, we see that there is another possibility—the upper lobe on C-1' might reach round to the upper lobe on C-1 **3.5**. This would clearly be a bonding interaction, and would allow concerted formation of both bonds. The problem, at least with molecules like ethylene, is that it is physically impossible for this overlap to develop at the same time as the overlap between C-2 and C-2' is maintained. The system is not flexible enough, and it only becomes flexible enough if one of the conjugated systems is long enough to allow a substantial amount of twisting (as drawn for the hypothetical case **2.85** in Chapter 2 on p. 20). We shall see later, in Section 3.6 on p. 49, one example where this might actually happen.

Frontier orbitals also explain why the rules change so completely for photochemical reactions. In a photochemical cycloaddition, one molecule has had one electron promoted from the HOMO to the LUMO, and this excited-state molecule reacts with a molecule in the ground state. The interacting orbitals that most effectively lower the energy of the transition structure are now the LUMO of the ground-state molecule and the orbital that had been the LUMO of the excited molecule, here called the 'LUMO'. The interaction of the HOMO with the 'HOMO' is likewise energy-lowering. LUMO/'LUMO' **3.6** and HOMO/'HOMO' **3.7** interactions are bonding at both ends in a [2 + 2] cycloaddition, and the two molecules can combine. This frontier orbital interaction is between orbitals close, or even identical, in energy, and so the two molecules are positively attracted to each other, giving an intermediate called an exciplex, instead of being minimally repelled, as in thermal cycloadditions. The interactions in a photochemical [4 + 2] cycloaddition **3.8** and **3.9** are antibonding at one end, and this type of cycloaddition cannot be concerted.

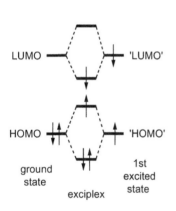

3.5

ground state — exciplex — 1st excited state

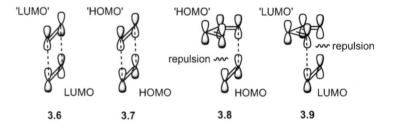

3.6   **3.7**   **3.8**   **3.9**

In contrast, applying frontier orbital theory to unimolecular reactions like electrocyclic reactions and sigmatropic rearrangements is inherently contrived, since we have artificially to treat a single molecule as having separate components, in order to have any frontier orbital interaction at all. Furthermore, frontier orbital theory does not explain why the barrier to forbidden reactions is so high—whenever it has been measured, the transition structure for the forbidden pathway has been 40 kJ mol^{-1} or more above that for the allowed pathway.

Frontier orbital theory is much better at dealing with small differences in reactivity, and we shall return to it in order to explain the much weaker elements of selectivity, like the effect of substituents on the rates and regioselectivity, and the *endo* rule, but we must look for something better to explain why interacting molecules conform to the rules with such dedication.

3.5 Correlation diagrams

Correlation diagrams provide a compelling explanation, at least for those reactions that have well-defined elements of symmetry preserved throughout the reaction. The idea is to identify the symmetry elements, classify the orbitals undergoing change with respect to those symmetry elements, and then see how the orbitals of the starting materials change into those of the product. The assumption is that an orbital in the starting material must feed into an orbital of the same symmetry in the product. Substituents, whether they technically break the symmetry or not, are treated as insignificant perturbations on the orbitals actually undergoing change. We shall now go through the sequence of steps involved in setting up a correlation diagram. It may be some comfort to know that it will not prove to be necessary to go to this much trouble for every new reaction you come across—it is a revealing exercise, and good practice in handling molecular orbitals, but you will be able to dispense with it as a day-to-day tool when you learn the Woodward–Hoffmann rules that came out of it.

We shall begin with an allowed reaction, the ubiquitous Diels–Alder.

a plane of symmetry intersects the page here

3.10

Step 1. Draw the bare bones of the reaction **3.10**, and put in the curly arrows for the forward and backward reactions. Remember that any substituents, even if they make the diene or dienophile unsymmetrical, do not fundamentally disturb the symmetry of the orbitals directly involved.

Step 2. Identify the orbitals undergoing change—the curly arrows help you to focus on what they are. For the starting materials, they are the π-orbitals (ψ_1-ψ_4^*) of the diene unit and the π-orbitals (π and π^*) of the C=C double bond of the dienophile. For the product, they are the π-bond and the two newly formed σ-bonds.

Step 3. Identify any symmetry elements maintained throughout the course of the reaction. There may be more than one. For a Diels–Alder reaction, which

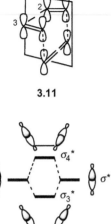

3.11

we know to be suprafacial on both components **3.11**, there is only the one, a plane of symmetry bisecting the bond between C-2 and C-3 of the diene and the π-bond of the dienophile.

Step 4. Rank the orbitals approximately by their energy, and draw them as energy levels, one above the other, with the starting material on the left and the product on the right (Fig. 3.1).

Step 5. Beside each energy level, draw the orbitals, showing the signs of the coefficients of the atomic orbitals. All the π-bonds are straightforward, but we meet a problem with the two σ-bonds in the product, which appear at first sight to be independent entities. In the next step we are going to have to identify the symmetry these orbitals have with respect to the plane of symmetry maintained through the reaction, and it is not obvious how to do this for a pair of independent-seeming orbitals. The answer is to combine them. They are, after all, held one bond apart, and they must interact quite strongly in a π sense. The interaction of the two bonding σ and the two antibonding σ* orbitals leads to a new set of four molecular orbitals σ_1, σ_2, σ_3^*, and σ_4^*, one pair (σ_1 and σ_3^*) lowered in energy because of the extra π-bonding, and the other pair (σ_2 and σ_4^*) raised in energy because of the extra π-antibonding. The interaction is overall mildly energy raising, as is usual with the interaction of filled orbitals with filled orbitals, and should be familiar as the energetic penalty in conformational analysis when two substituents eclipse each other.

Step 6. Classify each of the orbitals with respect to the symmetry element. Starting at the bottom left of Fig. 3.1, the lowest energy orbital is ψ_1 of the diene, with all-positive coefficients in the atomic orbitals, in other words with unshaded orbitals across the top surface of the conjugated system. The atomic orbitals on C-1 and C-2 are reflected in the mirror plane, intersecting the page at the dashed line, by the atomic orbitals on C-3 and C-4, and ψ_1 is therefore classified as symmetric (S). Moving up the left-hand column, the next orbital is the π-bond of the dienophile, which is also symmetric with respect to reflection in the plane. The next orbital is ψ_2 of the diene, in which the atomic orbitals on C-1 and C-2 have positive coefficients, and those on C-3 and C-4 have negative coefficients, because of the node half way between C-2 and C-3. The atomic orbitals on C-1 and C-2 are not reflected by the orbitals on C-3 and C-4, and this orbital is antisymmetric (A) with respect to the mirror plane. It is unnecessary to be any more sophisticated in the description of symmetry than this. The remaining orbitals can all be classified similarly as symmetric or antisymmetric. Likewise with the orbitals of the product on the right, σ_1 is symmetric, σ_2 antisymmetric, and so on.

Step 7. Complete the orbital correlation diagram (Fig. 3.1). Following the assumption that an orbital in the starting material must feed into an orbital of the same symmetry in the product, draw lines connecting the orbitals of the starting materials to those of the products nearest in energy and of the same symmetry. Thus, ψ_1 (S) connects to σ_1 (S), π (S) connects to π (S), and ψ_2 (A) connects to σ_2 (A), and similarly, with the unoccupied orbitals, ψ_3^* (S) connects to σ_3^*, π^* (A) connects to π^* (A), and ψ_4^* (A) connects to σ_4^* (A).

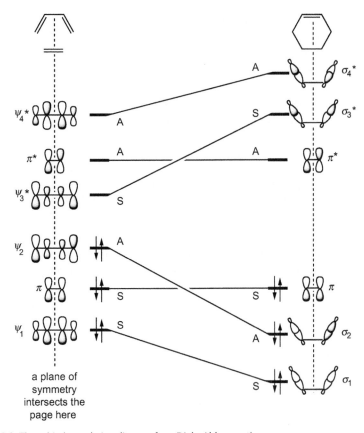

Fig. 3.1 The orbital correlation diagram for a Diels–Alder reaction

Step 8. Construct a state correlation diagram. In Fig. 3.1 the orbitals of the ground state in the starting material move smoothly into the orbitals of the ground state in the product. The ground state of the starting materials is designated ($\psi_1^2\pi^2\psi_2^2$). Because all the terms are squared (each of the orbitals is doubly occupied), it is also described as overall symmetric (S). Similarly the ground state of the product is ($\sigma_1^2\sigma_2^2\pi^2$), and it too is symmetric. The state correlation diagram is correspondingly easy, at least as far as we need to take it. Because the individual orbitals of the ground state in the starting material correlate with the individual orbitals of the ground state of the product, the state correlation diagram consists simply of a line joining the ground state with the ground state (Fig. 3.2).

Fig. 3.2 Part of the state correlation diagram for a Diels–Alder reaction

The state correlation diagram does not seem to have been very necessary here, but it is revealing when we build one for the forbidden thermal [2 + 2] cycloaddition like the dimerization of ethylene **3.12**, going through the same eight steps.

We draw the reaction and put in the curly arrows—the components are evidently the two π-bonds. There are two symmetry elements maintained this time—a plane like that in the Diels–Alder reaction, bisecting the π-bonds, but also another between the two reagents, which reflect each other through that plane.

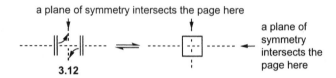

3.12

In order to classify the symmetry of the orbitals with respect to that plane, we have to take the approaching π-bonds and pair them up in a higher energy antisymmetric **3.13** and a lower energy symmetric **3.14** combination, which you can imagine as the molecular orbitals developing when the two molecules approach each other. It is essentially the same as the interaction of the σ-bonds that we used in setting up σ_1-σ_4* in the Diels–Alder reaction. We shall also have to repeat that exercise in this case, to deal with the two σ-bonds in the cyclobutane product.

We can now construct the orbital correlation diagram Fig. 3.3, but we must classify the symmetry of the orbitals twice over, once for the plane bisecting the

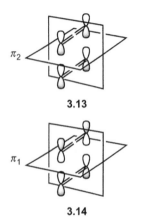

3.13

3.14

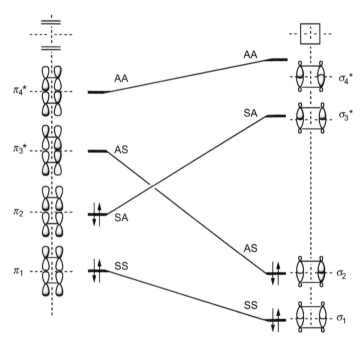

Fig. 3.3 The orbital correlation diagram for a 2 + 2 cycloaddition

π-bonds, represented by the vertical dashed line in Fig. 3.3, and then for the plane between the two reagents, the horizontal dashed line.

Thus, the lowest energy orbital in the starting materials is the bonding combination π_1 of the two bonding π-orbitals. This orbital is reflected through both planes and is classified as symmetric with respect to both (SS). The next orbital up is the antibonding combination π_2 of the two bonding π-orbitals. This orbital is reflected through the first plane, but not in the second, so it is classified as symmetric with respect to one and antisymmetric with respect to the other (SA). Work your way up through the two antibonding π-orbitals to see that π_3^* and π_4^* are AS and AA, respectively. The product side is similar—except for the addition of the second symmetry classification, it reproduces the pattern for the σ-bonds that we saw in Fig. 3.1.

We can now complete Fig. 3.3 by correlating the energy levels, feeding the orbitals in the starting materials into orbitals of the same symmetry in the product, SS to SS, SA to SA, AS to AS, and AA to AA. This time, the filled, bonding orbitals of the starting materials, π_1 and π_2, do not lead to the ground state orbitals of the product—one of them, π_1, leads to the lower bonding orbital σ_1, but the other, π_2, leads to one of the antibonding orbitals σ_3^*.

In the state correlation diagram (Fig. 3.4), the ground state of the starting materials, $\pi_1^2\pi_2^2$, is overall symmetric, because both terms are squared. Following the lines across Fig. 3.3, we see that this state feeds into a doubly excited state, $\sigma_1^2\sigma_3^{*2}$, in the product, which is also symmetric because both terms are squared. If we now start at the ground state of the product, $\sigma_1^2\sigma_2^2$, and follow the lines (SS and AS) in Fig. 3.3 back to the orbitals of the starting material, we find another doubly excited state $\pi_1^2\pi_3^{*2}$. Both of these states, with both terms squared, are again symmetric.

Any hypothetical attempt by the molecules to follow these paths in either direction, supposing they had the very large amounts of energy necessary to do so, would be thwarted, because states of the same symmetry cannot cross. The hypothetical reaction would, in fact, lead from ground state to ground state, but it would have to traverse a very substantial barrier, represented on Fig. 3.4 by the line E. This barrier provides, at last, a convincing explanation of why the forbidden [2 + 2] cycloaddition is so difficult—the energy needed to surmount it

E is the symmetry-imposed barrier to the concerted cycloaddition.

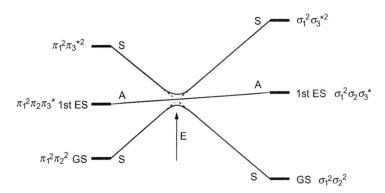

Fig. 3.4 The state correlation diagram for a [2 + 2] cycloaddition

is of the order of electronic excitation energies, far above that available in most thermal reactions.

We look now at the first excited state, $\pi_1{}^2\pi_2\pi_3{}^*$, which is produced by promoting one electron from π_2 to $\pi_3{}^*$. Following the lines in Fig. 3.3 from the occupied and the two half-occupied orbitals on the left (SS, SA, and AS), we are led to the orbitals of the first excited state on the right, $\sigma_1{}^2\sigma_2\sigma_3{}^*$. In the state correlation diagram, Fig. 3.4, both of these states are antisymmetric, and there is a line joining them, passing close to the avoided crossing in the ground state correlation. The value of E is evidently high, because it is approaching the energy of electronic excitation. It also explains why the photochemical [2 + 2] reaction is allowed—the electrons in the orbitals of the first excited state move smoothly over into the orbitals of the first excited state of the product. This does not mean that the reaction ends there, for the electron in $\sigma_3{}^*$ must somehow drop into σ_2 to give the ground state, disposing of a large amount of energy—by no means a simple event. All we need to understand in the present context is that the photochemical [2 + 2] reaction does not meet a symmetry-imposed barrier like that for the ground-state reaction.

Correlation diagrams have given us a convincing sense of where the barriers come from for those reactions that we have been calling forbidden. In principle, of course, no reaction is forbidden—what these reactions have is a formidable symmetry-imposed barrier, and something very unusual is needed if barriers of this magnitude are to be crossed. However, correlation diagrams take quite a bit of thought, and there are some pitfalls in their construction—however satisfying they may be, they are not for everyday use.

3.6 The Woodward–Hoffmann rules applied to cycloadditions

Fortunately, all the conclusions that can be drawn laboriously from correlation diagrams can be drawn more easily from a pair of rules, known as the **Woodward–Hoffmann rules**, which distil the essence of the idea into two statements governing all pericyclic reactions, one rule for thermal reactions and its opposite for photochemical reactions. Correlation diagrams explain why they work, but we are no longer dependent upon having to construct such diagrams. We shall examine the rules in this chapter with application to the cycloadditions that we have already seen in the last chapter, and then we shall see how they work for the other kinds of pericyclic reaction in the remaining chapters. The rule for thermal reactions is:

The Woodward–Hoffmann rule for thermal pericyclic reactions:

A ground-state pericyclic change is symmetry-allowed when the total number of (4q+2)ₛ and (4r)ₐ components is odd.

The **components** of a cycloaddition are obvious enough—we have been using the word already in this chapter to refer to the core electronic systems undergoing change. For a Diels–Alder reaction the components are the π-orbitals of the diene, containing four electrons, and the π-bond of the dienophile, containing two. As usual, we ignore all substituents not directly involved. What we have to do is to ask ourselves two questions: (1) which of these components is acting in a suprafacial manner and which in an antarafacial manner, and (2) in which of these components can the number of electrons be expressed in the form $(4q + 2)$ and in which in the form $(4r)$, where q and r are integers? For the Diels–Alder reaction, as we know, both components are undergoing bond formation in a suprafacial sense, as shown by the dashed lines in **3.15**, and so the answer to the first question is: both components.

The diene has four π-electrons, a number that can be expressed in the form $(4r)$, with r = 1. Since the new bonds are forming on the diene in a suprafacial manner, both lines coming to the lower surface, the diene is a $(4r)_s$ component. The dienophile has two electrons, a number that can be expressed in the form $(4q + 2)$, with q = 0. Since the new bonds are forming on the dienophile in a suprafacial manner, both lines coming to the upper surface, the dienophile is a $(4q + 2)_s$ component. Thus, we have put the answers to the two questions together, by asking ourselves: how many of the $(4q + 2)$ components are suprafacial, and how many of the $(4r)$ components are antarafacial? In the Diels–Alder reaction there is one $(4q + 2)_s$ component and no $(4r)_a$ components. We ignore $(4q + 2)_a$ and $(4r)_s$ components, when there are any. The total number of $(4q + 2)_s$ and $(4r)_a$ components is therefore 1, and, since this is an odd number, the reaction is symmetry-allowed.

The Diels–Alder reaction **3.15** is, as we have been calling it all along, a $[4 + 2]$ cycloaddition. Since it takes place suprafacially on both components, it is more informatively described as a $[4_s + 2_s]$ cycloaddition, and finally, because both components are π-systems, it is fully described as a $[_\pi4_s + _\pi2_s]$ cycloaddition. The labels $_\pi4_s$ and $_\pi2_s$ can be placed beside the appropriate component in the drawing, as in **3.15**, to help to identify what is going on, and the sum can conveniently be placed near the drawing, to show that the full check on the allowedness of the reaction has been completed.

The description $[_\pi4_s + _\pi2_s]$ for a Diels–Alder reaction does not supplant the older name—it is not the only reaction that is $[_\pi4_s + _\pi2_s]$, and so the name Diels–Alder is still needed to describe the reaction. 1,3-Dipolar cycloadditions **3.16** are equally $[_\pi4_s + _\pi2_s]$, and so are the combinations: allyl anion and alkene, allyl cation and diene, and pentadienyl cation and alkene. Furthermore, $[_\pi4_s + _\pi2_s]$ is not the only way of describing a Diels–Alder reaction. It would be easy to overlook the fact that the diene can be treated as one component, and to see it instead as two independent π-bonds. Although it makes extra work to see it this way, it does not cause the rule to break down. For example, the drawing **3.17** might have been used. The dashed line representing the developing overlap for the formation of the π-bond is from the lower lobe on C-2 to the lower lobe on C-3. This makes all three components suprafacial—the π-bond between C-1 and C-2 has both dashed lines to the lower lobes, and the π-bond between C-3 and C-4 also has both dashed lines to the lower lobes. In other words both are seen as suffering suprafacial development of overlap. The same is true for the π-bond of the dienophile. Overall the little sum is changed to having three $(4q + 2)_s$ components, which is still an odd number,

What do the terms $(4q + 2)_s$ and $(4r)_a$ mean?

$_\pi4_s$

$_\pi2_s$

3.15

No. of $(4q + 2)_s$ components: 1
No. of $(4r)_a$ components: 0
Total: 1
Odd number, allowed   ✓

$_\pi4_s$

$_\pi2_s$

3.16

and so the reaction remains allowed. It is now described as a $[_{\pi}2_s + _{\pi}2_s + _{\pi}2_s]$ cycloaddition.

Some other ways of describing a
Diels–Alder reaction

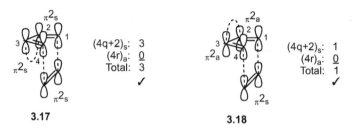

3.17   **3.18**

Another drawing **3.18**, still representing exactly the same reaction, places the dashed line between the upper lobes on C-2 and C-3. This changes each of the π-bonds of the diene to be seen as suffering notional antarafacial development of overlap. It is just as valid a representation as either of the earlier versions, and the sum still comes out with an odd number of $(4q + 2)_s$ components and no $(4r)_a$ components. The two $_{\pi}2_a$ components have two electrons each, which is a $(4q + 2)$ number and $(4q + 2)_a$ components do not have to be counted. The reaction is now a $[_{\pi}2_s + _{\pi}2_a + _{\pi}2_a]$ cycloaddition. Clearly, the three designations $[_{\pi}4_s + _{\pi}2_s]$, $[_{\pi}2_s + _{\pi}2_s + _{\pi}2_s]$, and $[_{\pi}2_s + _{\pi}2_a + _{\pi}2_a]$ are all the same reaction, and none of them defines a Diels–Alder reaction. The three designations, in fact, define where the dashed lines have been drawn in the three drawings **3.15**, **3.17**, and **3.18**, and no reaction should be described in this way in the absence of a drawing like these. We might note here that the words suprafacial and antarafacial are being used in the same sense as that in which they were defined in Fig. 2.7 in Chapter 2 on p. 20, but to refer to a drawing rather than a reaction.

A fourth of many possible drawings, **3.19**, is a *different* reaction, with the dashed line on the left at the back indicating that the upper lobe on C-2 is turning downwards to overlap with the lower lobe on C-3. This is not what happens—if it were to happen, it would produce a *trans* double bond in the cyclohexene product. Not only would that be impossibly strained, it is also forbidden, as the sum shows—there are two $(4q + 2)_s$ components. We see from the three drawings **3.15**, **3.17**, and **3.18** that there is considerable latitude in how to place the dashed lines to identify developing overlap, but they must identify the overlap that is actually developing, just as all three drawings correctly have the suprafacial attack on the π-bonds where the two σ-bonds are forming, and a π-bond forming between C-2 and C-3 without twisting.

A few words of warning about some other common confusions:

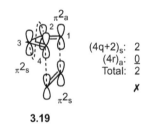

3.19

- The labels $_{\pi}4_s$ and $_{\pi}2_s$, apply to the *components*; be sure to focus on them, and not just on adjacent p-orbitals, before you check whether the dashed lines coming to the lobes at each end identify them as suprafacial or antarafacial. It is a common mistake to look at each end of a dashed *line*, such as the line between the upper lobes on C-2 and C-3 in the drawing **3.18** (repeated on the left), and call it suprafacial. If this is where you are looking, you need to recognize that the connection between C-2 and C-3 is

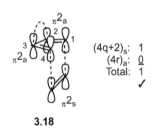

3.18

not a component. The components are the π-bonds between C-1 and C-2 and between C-3 and C-4. Placing the labels close to the components they refer to, and using the longest conjugated system as one component, help to avoid this confusion.

- The numbers of electrons in the components of a $[_\pi4_s + _\pi2_s]$ reaction are the key to getting the right answer; the 4 and the 2 are printed full size, and on the line. The s and the a, identifying whether they are suprafacial or antarafacial, and the π and σ, which we shall come to shortly, are qualifiers, which are printed as small subscripts. There is a tendency for the qualifiers to grow orthographically, and you will find in your reading designations like $[\pi4_s + \pi2_s]$, with many variants, perversely emphasizing the least important feature. Remember that it is first and foremost a [4 + 2] cycloaddition.

- When working out whether or not a reaction is obeying the rule, it is inappropriate to shade the orbitals—this is not a frontier orbital treatment, and no particular orbital is being considered when an analysis like **3.15** is carried out. There is no real need to draw the lobes at all, as long as the dashed lines clearly identify on which side of the atoms the new bonds are developing.

- It is more than half the battle if you make a good drawing, realistically representing a transition structure, and showing overlap that looks as though it could actually develop into a bond.

So far we have only defined what suprafacial and antarafacial mean on π-systems (Fig. 2.7), but we need to see how σ-bonds are treated by the Woodward–Hoffmann rules. Just as a suprafacial event on a π-bond has overlap developing to the two upper lobes that contribute to the bond, so with σ-bonds (Fig. 3.5a), overlap that develops to the two large lobes of the sp^3 hybrids is suprafacial. Less obviously, overlap that develops to the two small lobes is also suprafacial, because it is the counterpart to overlap developing to both lower lobes in a π-bond. Antarafacial overlap (Fig. 3.5b) is when one bond is forming to a large lobe and one to a small lobe, either way round.

A retro-Diels–Alder reaction **3.20** will allow us to look at an example involving single bonds. We have three components: two σ-bonds and one π-bond. Let us just do two, **3.21** and **3.22**, of the several possible equivalent ways of drawing this reaction. Notice that the dashed lines correspond to the stereochemistry we must conform to—the experimentally observed counterpart to a suprafacial reaction on both components in the normal direction.

3.20

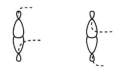

   (a) Suprafacial bond formation   (b) Antarafacial bond formation

Fig. 3.5 Suprafacial and antarafacial, defined for σ-bonds

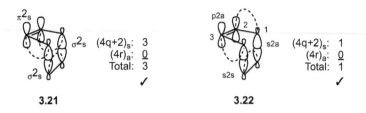

3.21   **3.22**

Version **3.21** uses only suprafacial components: the π-bond has both lines coming from the bottom lobes, and both σ-bonds have the dashed lines coming from the large lobes. In this picture it is a $[_\pi2_s + _\sigma2_s + _\sigma2_s]$ cycloreversion. The alternative **3.22** arbitrarily has the dashed line representing the developing overlap between C-1 and C-2 on the top surface. This causes one line to the π-bond to come into the lower lobe on C-3, as before, but the other now comes into the upper lobe on C-2, making the π-bond a $_\pi2_a$ component. The σ-bond at the front is unchanged as $_\sigma2_s$, but the σ-bond at the back has one dashed line coming into the large lobe and the other into the small lobe making it $_\sigma2_a$. In this picture it is a $[_\pi2_a + _\sigma2_s + _\sigma2_a]$ cycloreversion, and is, of course, equally allowed. As before, the designation does not define the reaction, it defines the drawing.

Here are some longer conjugated systems undergoing cycloadditions: the cycloaddition **2.66** in Chapter 2 on p. 16 of a pentadienyl cation to an alkene can be drawn as the $[_\pi4_s + _\pi2_s]$ cycloaddition **3.23**, and the [6 + 4] cycloaddition of tropone **2.76** to cyclopentadiene in Chapter 2 on p. 17 can be fleshed out now as the $[_\pi6_s + _\pi4_s]$ cycloaddition **3.24**.

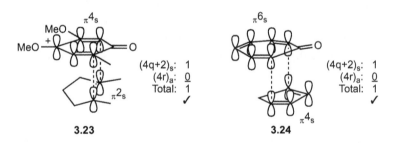

3.23   **3.24**

A short-cut rule, to be applied with caution

It is generally a good idea to try to draw the dashed lines to make as many suprafacial components as you can, for it is possible to simplify the rule further when you do. If the *total* number of electrons involved is a (4n + 2) number, the all-suprafacial reaction will be allowed, as in **3.21**. This can be made to apply to a high proportion of the pericyclic reactions you will ever come across, and especially to cycloadditions, as we saw with a preliminary version of this rule in the last chapter. On the other hand, if the total number of electrons is a 4n number, the reaction will be allowed if one, and only one, component is antarafacial. By choosing to place the dashed lines to make as many suprafacial components as you can, this rule can be made to apply to very nearly all the remaining pericyclic reactions. These versions of the general rule allow one to come quickly to a conclusion about whether a particular reaction will prove to be allowed or not. Then, if the reaction is at all complicated, it can be checked against the full rule. Most of the reactions analysed in the rest of this book will use suprafacial components wherever that is possible, so that the antarafacial components will stand out.

Perhaps it stands out best in this, one of the most spectacular achievements of the theory of pericyclic reactions. Tetracyanoethylene adds to heptafulvalene in a [14 + 2] cycloaddition **3.25**. Because the number of electrons involved is 16 (eight curly arrows), this reaction appears at first sight to be forbidden—16 is an antiaromatic number, since it can be expressed in the form 4n. We can now see that it will be allowed if one of the two components can react in an antarafacial manner, and the heptafulvalene is evidently flexible enough not to lose the conjugation through the seven double bonds while allowing overlap to develop onto opposite sides of the conjugated system, making it an antarafacial component. The two hydrogen atoms in the adduct **3.27** prove to be *trans* to each other, showing that the heptafulvalene has suffered antarafacial attack **3.26**: the dashed lines show the overlap developing, one to a lower lobe and one to an upper, making the heptafulvalene a $_\pi 14_a$ component, and the reaction an allowed $[_\pi 14_a + _\pi 2_s]$ cycloaddition.

$(4q+2)_s$:	1
$(4r)_a$:	0
Total:	1

3.25   3.26   3.27

The rule for photochemical reactions is simply the reverse of the rule for thermal reactions:

A pericyclic change in the first electronically excited state is symmetry-allowed when the total number of $(4q+2)_s$ and $(4r)_a$ components is even.

The Woodward–Hoffmann rule for photochemical pericyclic reactions

It is not clear how many photochemical reactions are in fact pericyclic, and it is not easy to use the usual methods of physical organic chemistry to prove it. What is clear is that a great many photochemical reactions do obey the photochemical rule, and it therefore seems likely that some of them have pericyclic character at a critical stage. Let us look at two possible geometries for [2 + 2] cycloadditions. Both components can act suprafacially **3.28**, or one can be suprafacial and one antarafacial **3.29**. The first **3.28** is a $[_\pi 2_s + _\pi 2_s]$ cycloaddition with an even number of $(4q + 2)_s$ components and no $(4r)_a$ components. It is symmetry-forbidden as a thermal reaction, but symmetry-allowed photochemically. The second **3.29** is a $[_\pi 2_s + _\pi 2_a]$ cycloaddition, with one $(4q + 2)_s$ component making it thermally allowed; but its orbitals cannot reach each other.

$\pi 2_s$

$(4q+2)_s$:	2
$(4r)_a$:	$\underline{0}$
Total:	2
Thermal:	✗
Photo:	✓

$\pi 2_s$

3.28

$\pi 2_a$

$(4q+2)_s$:	1
$(4r)_a$:	$\underline{0}$
Total:	1
Thermal:	✓ (but unreasonable)
Photo:	✗

$\pi 2_s$

3.29

With the pattern for [2 + 2] cycloadditions fresh in mind, we turn to the anomalous reactions that appear to be forbidden [2 + 2] cycloadditions taking place thermally under mild conditions.

3.7 Some anomalous [2 + 2] cycloadditions

One group of anomalous reactions is that of ketenes with alkenes. These reactions appear to have some of the characteristics of pericyclic cycloadditions, such as being stereospecifically *syn* with respect to the double bond geometry (i.e. suprafacial at least on the one component), as in the reactions of the stereoisomeric cyclooctenes **3.30** and **3.32** giving the diastereoisomeric cyclobutanones **3.31** and **3.33**.

3.30   **3.31**   **3.32**   **3.33**

On the other hand, stereospecificity is not always complete, and many ketene cycloadditions take place only when there is a strong donor substituent on the alkene. An ionic stepwise pathway by way of an intermediate zwitterion **3.34** is therefore entirely reasonable in accounting for many ketene cycloadditions. It seems likely that some of these reactions are pericyclic and some not, with the possibility of there being a rather blurred borderline between the two mechanisms, with one bond forming so far ahead of the other that any symmetry in the orbitals is essentially lost. But we are still left with the question: when it is pericyclic, how does it overcome the symmetry-imposed barrier?

One suggestion is that the two molecules approach each other at right angles, with overlap developing **3.35** in an antarafacial sense on the ketene, making the reaction the allowed $[_\pi 2_s + _\pi 2_a]$ cycloaddition that we have dismissed as being unreasonable. This is the most simple explanation, but it is unsatisfactory because other alkenes and alkynes might be able to achieve a transition structure with about the same degree of steric hindrance in the approach of the two π-bonds as this, yet such cycloadditions do not occur.

The probability of some [2 + 2] cycloadditions of ketenes being concerted is more likely to be a consequence of the fact that ketenes have two sets of π-orbitals

3.34

$\pi 2_s$ O $\pi 2_a$

3.35

at right angles to each other. Overlap can develop to orthogonal orbitals **3.36** (dashed lines), and in addition there is transmission of information from one orbital to its orthogonal neighbour (heavy line). This is a legitimate but somewhat contrived way of making the electronic connection cyclic and hence pericyclic. This version identifies the reaction as an allowed $[_{\pi}2_s + _{\pi}2_a + _{\pi}2_a]$ cycloaddition. In essence the ketene is able to take up the role of antarafacial component by using an orbital that has turned through 90° towards the alkene component. A variant of this approach, perhaps the easiest way of thinking about these reactions, is to omit the overlap drawn with a heavy line in **3.36**, and to concentrate on the key, σ-bond-forming events. This removes the symmetry-imposed barrier, because the reaction is no longer being thought of as strictly pericyclic. The two bonds are still being formed more or less in concert, but independently. The symmetry information is not transmitted from one orbital to the other, because they are orthogonal.

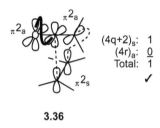

The equivalent frontier orbital treatment satisfyingly shows that the bond forming between C-1 and C-1′ develops mainly from the interaction of the p-orbital on the carbonyl carbon from the LUMO of the ketene (π^* of the C = O group) and one lobe from the HOMO of the alkene **3.37**, and that the bond between C-2 and C-2′ develops mainly from the interaction of one lobe from the HOMO of the ketene (ψ_2 of the three-atom linear set of orbitals analogous to the allyl anion) and one lobe from the LUMO of the alkene **3.38**.

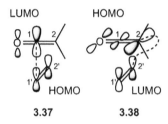

Related to ketene cycloadditions are the group of cycloadditions with vinyl cation intermediates, like the cation **2.170** in Chapter 2 on p. 30 in the Smirnov-Zamkov reaction. Another example is the reaction of allene with hydrogen chloride, taking place by way of the cycloaddition of the vinyl cation **3.39** to another molecule of allene to give the cyclobutyl cation **3.40**. Vinyl cations, like ketenes, have two p-orbitals at right angles to each other, and overlap can develop to each simultaneously **3.41**, just as it did with ketenes. In a sense, a ketene is merely a special case of a vinyl cation, with the carbonyl group a highly stabilized carbocation.

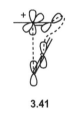

There are also several reactions in organometallic chemistry, which at first sight appear to contravene the rule but which can be explained in a somewhat similar way. Hydrometallation, carbometallation, metallo-metallation, and olefin metathesis reactions all have the feature of being stereospecifically suprafacial additions to an alkene or alkyne. We can take hydroboration **3.42** as an example, where the curly arrows show it as a [2 + 2] cycloaddition of a σ-bond to a π-bond, for which the all-suprafacial pathway is forbidden. It is known not to be stepwise, beginning with electrophilic attack by the boron atom, because electron-donating substituents on the alkene do not speed up the reaction anything like as much as they do when alkenes are attacked by cationic electrophiles—there must be some component of hydride addition more or less concerted with the electrophilic attack. The empty p-orbital on the boron is the electrophilic site and the s-orbital of the hydrogen atom in the B–H bond is the nucleophilic site. These orbitals are orthogonal, and so the cycloaddition is not properly pericyclic. The frontier orbital pictures **3.43** and **3.44** reinforce the perception that the two bonds are formed independently, with the former illustrating the electrophilic attack by the borane, and the latter the nucleophilic attack by the borane.

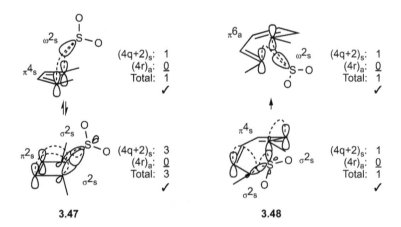

3.42   **3.43**   **3.44**

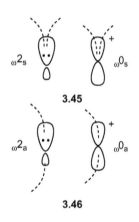

Another anomalous cycloaddition is the insertion of a carbene into an alkene. Six-electron cheletropic reactions (like the cycloreversions of **2.176** and **2.177** in Chapter 2 on p. 31) are straightforward allowed pericyclic reactions, which we can now classify with the two drawings **3.47**, one for the suprafacial addition of sulfur dioxide to the diene **2.179** and the other for its reverse. Similarly, we can draw **3.48** for the antarafacial addition of sulfur dioxide to the triene **2.180** and its reverse. The new feature here is that one of the orbitals is a lone pair, which is given the letter ω to distinguish it from σ- and π-bonds, with suprafacial and antarafacial defined by the drawings **3.45** and **3.46**, which apply to all p-orbitals and to sp^3 hybrids, filled or unfilled.

	$(4q+2)_s$:	1
	$(4r)_a$:	0
	Total:	1

	$(4q+2)_s$:	3
	$(4r)_a$:	0
	Total:	3

	$(4q+2)_s$:	1
	$(4r)_a$:	0
	Total:	1

	$(4q+2)_s$:	1
	$(4r)_a$:	0
	Total:	1

3.47   **3.48**

The problem comes with the insertion of a carbene into a double bond, which is well known to be stereospecifically suprafacial with singlet carbenes like dichlorocarbene (See Section 2.12, p. 30). This is clearly a forbidden pericyclic reaction, if it takes place in the sense **3.49** → **3.50**. This is known as the linear approach, in which the carbene, with its two substituents already lined up where they will be in the product, comes straight down into the middle of the double bond. The two reactions above, **3.47** and **3.48**, are also linear approaches, but these are both allowed, the former because the total number of electrons (6) is a $(4n + 2)$ number, and the latter because the triene is flexible enough to take up the role of antarafacial component.

3.49   **3.50**

The forbidden linear approach of a carbene to an alkene

The alternative is a non-linear approach, in which the carbene approaches the double bond on its side, and then has the two substituents tilt upwards as the reaction proceeds, in order to arrive in their proper orientation in the product **3.50**. This is best understood using the frontier orbitals **3.51a** and **3.51b**, where the HOMO of the carbene is the lone pair and the LUMO is the empty p-orbital. Each of these orbitals, when presented to the double bond with the carbene on its side, matches the signs of the appropriate atomic orbitals on the double bond (dashed lines).

HOMO   LUMO

LUMO   HOMO
3.51a   **3.51b**

3.50   **3.52**

The allowed non-linear approach of a carbene to an alkene

Once again, by separating the formation of the two σ-bonds into overlap developing independently to orthogonal orbitals on the carbene, we no longer need to see the reaction as strictly pericyclic. As with ketenes, however, it is possible to connect up the orthogonal orbitals, as in **3.52**, to make the non-linear approach classifiably pericyclic and allowed. This avoids any problem there might be with reactions like **3.47** and **3.48** being pericyclic and the clearly related reaction **3.51** seeming not to be.

3.8 Secondary effects

Although they are much less powerfully controlling than the constraints of orbital symmetry, there are some important secondary effects determining stereoselectivity, reaction rate, and regioselectivity. They are discussed here for cycloadditions, where they are most important and most easily explained.

Stereoselectivity in cycloadditions

One of the most challenging stereochemical findings is Alder's *endo* rule for Diels–Alder reactions (pp. 23 and 24). The favoured transition structures **2.110** and **2.113**, with the electron-withdrawing substituent in the more hindered environment, give the thermodynamically less favourable products **2.111** and **2.114** (see Chapter 2). Any reaction in which a kinetic effect overrides the usual thermodynamic effect on reaction rates is immediately interesting, and demands an explanation.

The simplest explanation comes from the frontier orbitals. As we shall see, they are the HOMO of the diene and the LUMO of the dienophile. The former is ψ_2 of butadiene, and the latter, using acrolein with four conjugated p-orbitals as a model, has a nodal pattern similar to that of the LUMO ψ_3^* of butadiene. If we place these orbitals in the appropriate places for the *endo* reaction **3.53**, we see that there is the usual primary interaction (dashed lines), consistent with the rules, as we saw in the drawing **3.3** in Section 3.4 on p. 38, but there is an additional bonding interaction (bold line) between the orbitals on C-2 of the diene and the carbonyl carbon of the dienophile. This interaction, known as a secondary orbital interaction, does not lead to a bond, but it does make a contribution to lowering the energy of this transition structure relative to that of the *exo* reaction, where it must be absent. It is equally compatible **3.54** with the reaction showing inverse electron demand **2.117** → **2.118** in Chapter 2 on p. 24, although there is some doubt whether that is actually pericyclic.

A secondary orbital interaction has been used to explain other puzzling features of selectivity, but, like frontier orbital theory itself, it has not stood the test of higher levels of theoretical investigation. Although still much cited, it does not appear to be the whole story, yet it remains the only simple explanation. It works for several other cycloadditions too, with the cyclopentadiene + tropone reaction favouring the extended transition structure **2.106** because the frontier orbitals have a repulsive interaction (wavy lines) between C-3, C-4, C-5, and C-6 on the tropone and C-2 and C-3 on the diene in the compressed transition structure **3.55**. Similarly, the allyl anion + alkene interaction **3.56** is a model for a 1,3-dipolar cycloaddition, which has no secondary orbital interaction between the HOMO of the anion, with a node on C-2, and the LUMO of the dipolarophile, and only has a favourable interaction between the LUMO of the anion and the HOMO of the dipolarophile **3.57**, which might explain the low level or absence of *endo* selectivity that dipolar cycloadditions show.

Substituent effects on rates of cycloadditions

The point was made early in the last chapter that Diels–Alder reactions take place faster if the dienophile has electron-withdrawing groups on it, and that electron-donating groups on the diene help too. Again the explanation comes most readily from the frontier orbitals. Put briefly, a donor substituent on the diene raises the energy of its HOMO, and an electron-withdrawing substituent

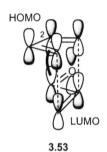

3.53

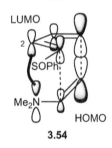

3.54

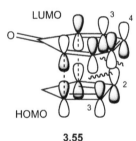

3.55

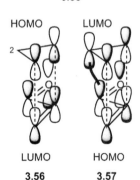

3.56   3.57

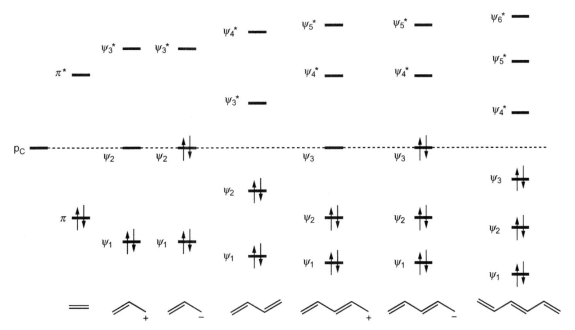

Fig. 3.6 The energies of the π molecular orbitals of the most simple conjugated systems

on the dienophile lowers the energy of its LUMO, bringing the two orbitals closer together in energy. Because orbitals close in energy interact more strongly than those far apart, the energy of the transition structure is lower than it was without the substituents.

It is possible to do better than just accept that the effects of electron-donating and withdrawing substituents will be this way round. A simple argument based on the molecular orbitals of conjugated systems allows us to deduce that donors raise the energy of molecular orbitals and that withdrawing groups lower them. Let us recall the energies of the molecular orbitals of the first six conjugated systems, starting with a p-orbital on carbon (one p orbital), which is used as a reference, and working up to hexatriene (six conjugated p-orbitals), with two entries for each of the ions, one for the cation and one for the anion, Fig. 3.6. We see immediately how the upper energy levels steadily rise and the lower steadily fall as more and more orbitals are brought into conjugation.

To estimate how the energies of the frontier orbitals change when an electron-donating or withdrawing substituent is added to a conjugated system we construct a diagram (Fig. 3.7) with the energy levels lifted from Fig. 3.6. The unsubstituted diene is on the far left of Fig. 3.7a and the unsubstituted dienophile is on the far right of Fig. 3.7b. To assess the effect of having an electron-donating group X on C-1 of the diene, we start with the best possible electron-donating group, which is a carbanion. Using the pentadienyl anion as a model for a 1-X-substituted diene, we place its energy levels on the right in Fig. 3.7a. We then assume that a less effective electron-donating substituent will make the orbitals of the conjugated system fall in between those of the unsubstituted diene

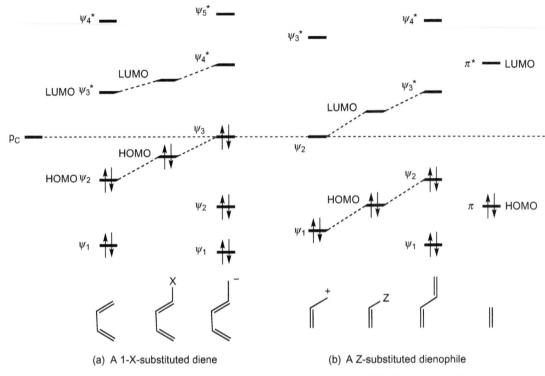

Fig. 3.7 Estimating the energy levels for the orbitals of (a) a 1-X-substituted diene and (b) a Z-substituted dienophile

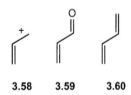

3.58   3.59   3.60

and of the pentadienyl anion. The energy levels of its frontier orbitals will then come between the corresponding energy levels for the two extreme versions, and we place these compromise levels in between the extremes as the middle column of Fig. 3.7a. We see immediately that the HOMO is higher in energy for the X-substituted diene than for the unsubstituted diene.

Estimating the effect of an electron-withdrawing group Z on an alkene is only a little more complicated. The argument begins with the idea that the best possible electron-withdrawing group would be a carbocation. Hence the allyl cation is the model for that extreme, and its energy levels can be placed on the left of Fig. 3.7b. Most electron-withdrawing groups, like carbonyl, nitrile and nitro, are not only electron-withdrawing, but also add further conjugation. The extreme version for such a group with no electron-withdrawing properties is best modelled by butadiene rather than by the unsubstituted alkene. Thus, acrolein **3.59** can be regarded as having properties somewhere between those of an allyl cation **3.58**, in which we disregard the oxyanion substituent completely, and butadiene **3.60**, in which we overestimate its contribution by disregarding the electronegativity of the oxygen atom. We therefore place the orbitals of butadiene as the third column in Fig. 3.7b, and predict, as before, that the frontier orbitals of a Z-substituted alkene will fall somewhere between the corresponding energy levels for the allyl cation and butadiene. We see that the energy of the LUMO of a Z-substituted alkene is very substantially lower than that of the unsubstituted alkene on the far right—some of the lowering in energy comes from the conjugation, with $\psi_3{}^*$ of butadiene already lower than π^*, and even more comes from its resemblance to the allyl cation. We can now see that the energy of the HOMO of the diene has indeed been raised and that of the LUMO of the dienophile lowered, and we no longer take it simply on faith.

Different electron-donating groups will make different contributions, with the more powerful, like oxyanions, increasing the proportion of pentadienyl anion character, and raising the energy of the HOMO even more. Different electron-withdrawing groups will also make different contributions, with the more powerful or more numerous increasing the proportion of allyl cation character and further lowering the LUMO energy. Lewis acid coordination to a carbonyl group will likewise decrease the influence of the oxygen lone pairs on the π-bond of the carbonyl group and make acrolein more like an allyl cation. This will lower the LUMO energy even more and explain the large rate accelerations found for Lewis acid-catalysed Diels–Alder reactions, such as that in the reaction of anthracene **3.61** with dimethyl fumarate **3.62**.

3.61   **3.62**

with AlCl$_3$ 2 h at r.t.
without AlCl$_3$ 2-3 days at 101°

Regioselectivity in cycloadditions

The usual pattern of regioselectivity in Diels–Alder reactions—the formation largely of 'ortho' and 'para' adducts—was outlined in Chapter 2 on pp. 25 and 26, where the inadequacy of an explanation based only on the charge distribution in the starting materials was mentioned. We shall find a better explanation in the frontier orbitals, where we look at the relative sizes of the coefficients on the individual atoms that are forming the new bonds.

We start with the unperturbed orbitals of the simple conjugated systems, analogous to Fig. 3.6. These are summarized in Fig. 3.8, where we are looking down on top of the p-orbitals, which are sized in proportion to the coefficients of the atomic orbitals in the molecular orbitals. The relative magnitudes of the numbers follow from the sinusoidal patterns of each successive orbital starting with the lowest in energy, with the limits for the sine curves set one bond length outside the outer atoms, as shown for the allyl system **3.63**. The absolute values then follow from the rule that the sum of the squares of all the atomic orbital coefficients in any one molecular orbital must add up to one.

To estimate how the HOMO of a 1-X-substituted diene is polarized, we merge ψ_2 of the unperturbed diene on the left of Fig. 3.9a with ψ_3 of the pentadienyl anion on the right of Fig. 3.9a. The X-substituted diene will come somewhere between these extremes, with a larger coefficient on C-4 than on C-1, but not with as big a difference between C-1 and C-4 as in the pentadienyl anion, which actually has a node, symbolized by a black dot, on the atom corresponding to C-1.

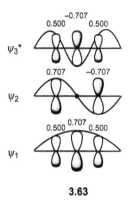

$\psi_3{}^*$ -0.707
            0.500   0.500

            0.707 -0.707
ψ_2

            0.500   0.707   0.500
ψ_1

3.63

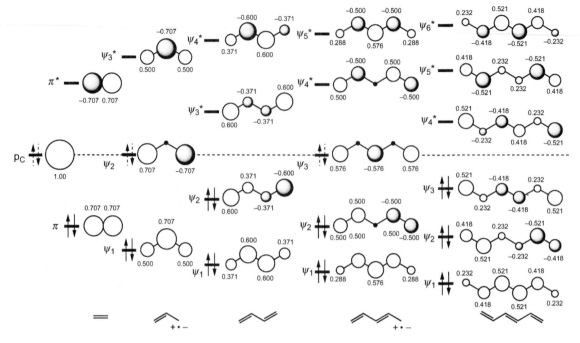

Fig. 3.8 Atomic orbital coefficients in the π molecular orbitals of the simplest conjugated systems

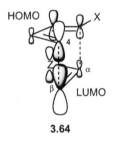

3.64

For the LUMO of a Z-substituted alkene, we combine ψ_2 of the allyl cation on the left of Fig. 3.9b with ψ_3^* of butadiene on the right. Both orbitals contribute to the β carbon of the alkene having the larger coefficient, just as they both contributed to lowering the LUMO energy. The last step in the argument is to recognize that better overlap develops if the two larger lobes approach each other than if the larger lobe of one component approaches the smaller lobe of the other. Thus C-4 of the diene bonds to the β carbon of the dienophile **3.64**. This picture

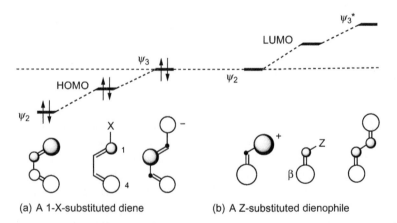

(a) A 1-X-substituted diene   (b) A Z-substituted dienophile

Fig. 3.9 Estimating the coefficients for (a) the HOMO of a 1-X-substituted diene and (b) the LUMO of a Z-substituted dienophile

is useful in representing something that was mentioned in Chapter 2 on pp. 27 and 28—pericyclic cycloadditions have two bonds forming at the same time, but that does not mean that they are formed to the same extent in the transition structure. The bond between C-4 of the diene and the β carbon of the dienophile forms ahead of the bond between C-1 of the diene and the α carbon of the dienophile, but the overlap leading to the latter is beginning. Asymmetry in the starting materials introduces a level of asymmetry in the transition structure, without necessarily making the reaction fully stepwise. In a highly unsymmetrical case, the benefit of having overlap develop between C-1 of the diene and the α carbon of the dienophile can become too small to be energetically profitable. When that happens, a stepwise pathway will be followed, even for a symmetry-allowed reaction, because a stepwise pathway does not suffer from the high entropy of activation associated with forming two bonds at the same time.

The language that has developed to deal with this subtlety is to use the word **concerted** when both bonds are forming at the same time, without necessarily implying that they are developing to an equal extent, and to use the word **synchronous** when they are developing to an equal extent. Most pericyclic reactions are therefore concerted but asynchronous, since most pericyclic reactions do not have symmetrical components. This terminology makes it easy to appreciate that there is a continuity between pericyclic and stepwise ionic reactions, with a blurred borderline between them.

So far our deductions about regiochemistry merely match the simple picture, **2.136** and **2.137** in Chapter 2 on p. 26, based on the total charge distribution in the starting materials. The real challenge is to explain the formation (p. 26) of the 'ortho' adduct **2.144** from the combination of pentadienoic acid **2.142**, a 1-Z-substituted diene, with acrylic acid **2.143**. The argument based on the total charge distribution does not work in this case, but the frontier orbital argument does. The LUMO of the dienophile is the same—it is repeated from Fig. 3.9b in Fig. 3.10b. What we need is to estimate how the HOMO of a 1-Z-substituted diene will be polarized. The extreme versions that we combine are the HOMO of hexatriene, ψ_3 shown on the left of Fig. 3.10a, and the HOMO of the pentadienyl cation, ψ_2, shown on the right. The latter is not polarized—its coefficients have the same absolute magnitude both on C-1 and on C-4. It is the element of conjugation from the hexatriene-like character that introduces a small polarization, with the coefficient on C-4 in ψ_3 marginally larger than that on C-1. The HOMO of the 1-Z-substituted diene is therefore polarized in the same sense as a 1-X-substituted diene, but to a much smaller degree, and the regiochemistry of its Diels–Alder reaction is explained.

The regioselectivity of 1,3-dipolar cycloadditions is not as simple to explain. The best explanation yet advanced is still based on frontier orbitals, but unfortunately it is no longer possible to deduce what they look like without computational support. Calculations have given representative sets of the energies and coefficients for the frontier orbitals of all the major dipoles. The coefficients have to be modified to take account of the fact that many dipoles have heteroatoms at their extremities, and bond-formation to heteroatoms requires a resonance integral different from the resonance integral for C–C bond-formation. From the energies it is possible to deduce whether the separation between the energy of

A distinction between the **concerted** formation of two bonds and the **synchronous** formation of two bonds.

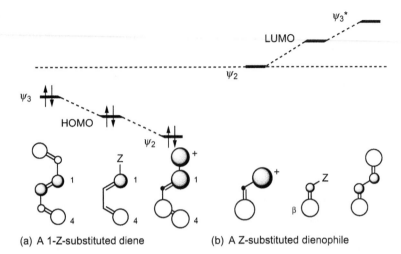

(a) A 1-Z-substituted diene   (b) A Z-substituted dienophile

Fig. 3.10 Estimating the coefficients for (a) the HOMO of a 1-Z-substituted diene and (b) the LUMO of a Z-substituted dienophile

the HOMO of the dipole and the LUMO of the dipolarophile is lower than that between the HOMO of the dipolarophile and the LUMO of the dipole. The former are called HO-dipole controlled, and the latter LU-dipole controlled.

The pattern that emerges is that dipoles calculated to have a high energy HOMO are the ones that react faster (or only) with dipolarophiles that have Z-substituents, and dipoles calculated to have a low energy LUMO are those that react faster (or only) with dipolarophiles that have X-substituents. In addition, the polarization deduced for the frontier orbitals matches to a large extent the regioselectivity observed in their reactions. The large number of dipoles, and the extended range created by the addition of Z- and/or X-substituents, make the subject too large to be covered any more fully here. It is enough to know that information is available if you ever need it.

3.9 **Further reading**

The application of molecular orbital theory to pericyclic reactions has been described, at a level similar to that here, in a number of textbooks:

T. L. Gilchrist and R. C. Storr, *Organic Reactions and Orbital Symmetry*, Cambridge University Press, 2nd Edn., 1972.

R. E. Lehr and A. P. Marchand, *Orbital Symmetry*, Academic Press, New York, 1972.

F. A. Carey and R. J. Sundberg, *Advanced Organic Chemistry*, Plenum, New York, 3rd Edn., 1990.

N. Isaacs, *Physical Organic Chemistry*, Longman, Harlow, 2nd Edn., 1995.

The 'bible', historically important, and a stimulating read, is:

R. B. Woodward and R. Hoffmann, *The Conservation of Orbital Symmetry*, Verlag Chemie, Weinheim, 1970.

The history of the early development of the ideas, in Woodward's own words, is Chapter 18 of *Robert Burns Woodward*, ed. O. T. Benfey and P. J. T. Morris, Chemical Heritage Foundation, Philadelphia, 2001.

There is a more extensive discussion in:

I. Fleming, *Molecular Orbitals and Organic Chemical Reactions—Student Edition*, Wiley, Chichester, 2009, and

I. Fleming, *Molecular Orbitals and Organic Chemical Reactions—Library Edition*, Wiley, Chichester, 2010.

Some specialized topics are covered in

T. T. Tidwell, *Ketenes*, Wiley, New York, 1995; R. E. Lehr and A. P. Marchand on criteria for concertedness in Ch. 1, Vol. 1, in *Pericyclic Reactions*, ed. A. P. Marchand and R. E. Lehr, Academic Press, New York, 1977; H. E. Zimmerman on Möbius-Hückel pericyclic transition structures, in Ch. 2, Vol. 1, in *Pericyclic Reactions* loc. cit.; W. M. Jones and U. H. Brinker on carbenes in Ch. 3, Vol. 1, *Pericyclic Reactions* loc. cit.; K. N. Houk on FMO theory of pericyclic reactions, in Ch. 5, Vol. 2, in *Pericyclic Reactions* loc. cit.; K. N. Houk and K. Yamaguchi on the theory of 1,3-dipolar cycloadditions in Ch. 12, in *1,3-Dipolar Cycloaddition Chemistry*, ed. A. Padwa, Vol. II, Wiley, New York, 1984.

3.10 **Problems**

3.1 Draw the frontier orbital interactions for the all-suprafacial cycloaddition of an allyl anion to an alkene and for an allyl cation to a diene showing that they match, and show that the alternatives, allyl cation with alkene and allyl anion with diene, are symmetry-forbidden.

3.2 Account for the change of product ratio in the following reaction as a result of the change of solvent from polar to non-polar (note that the solvents are inert in the conditions used for these reactions).

solvent: MeCN   9 : 1
              hexane 1 : 4

3.3 The following reactions take place with consecutive cycloadditions and retro-cycloadditions. Identify the steps, and show that both steps (and all three steps in the last example) obey the Woodward–Hoffmann rules by drawing realistic transition structures, adding the lines that identify the developing overlap and its stereochemistry (you will find coloured lines more satisfying than the dashed lines that have had to be used in this book), classifying the stereochemistry, suprafacial or antarafacial, for each component, and completing the sum calculating the total number of $(4q + 2)_s$ and $(4r)_a$ components:

3.4 Using the Woodward–Hoffmann rules, and realistic drawings of the transition structures, as in the preceding question, predict the stereochemistry of the double bonds produced in the following reactions, which involve stereospecific cycloreversions:

3.5 Identify any pericyclic cycloadditions or cycloreversions that take place within these multistep sequences:

3.11 **Summary**

- There is much evidence for the existence of concerted cycloadditions, in which both bonds are formed at the same time.

- There are three levels of explanation why some pericyclic cycloadditions are allowed while others are forbidden. The most simple is based on the aromaticity of the transition structure. The next most simple matches the signs of the atomic orbital coefficients in the frontier orbitals. (In Diels–Alder reactions the appropriate frontier orbitals are usually the HOMO of the diene and the LUMO of the dienophile.) The most thorough and satisfying explanation uses correlation diagrams in which first the orbitals and then the states of the reactants are connected to those of the cycloadducts. In particular the third explanation identifies why the energy barrier to forbidden reactions is so substantial.

- Nevertheless, if only *one* bond in a cycloaddition forms at a time, cycloadditions can take place that would be forbidden if both bonds were being formed at the same time. These cycloadditions are not pericyclic.

- A simple rule governs all thermal pericyclic reactions: A ground-state pericyclic change is symmetry-allowed when the total number of $(4q + 2)_s$ and $(4r)_a$ components is odd.

- Most pericyclic cycloadditions mobilize a total of $(4n + 2)$ electrons; they can then be described as all-suprafacial.

- If, on the other hand, and rarely, the total number of electrons mobilized is a number that can be described in the form $4n$, a pericyclic cycloaddition is symmetry-allowed if one of the components reacts in an antarafacial manner.

- Ketenes and carbenes, which appear at first sight to undergo forbidden cycloaddditions, are able to form both bonds concertedly. The most simple explanation is that orthogonal orbitals are involved, and the reaction is not then strictly pericyclic.

- The *endo* selectivity of Diels–Alder reactions can be explained by secondary overlap within the frontier orbital picture.

- The effects of substituents on rates and regioselectivity are most simply explained by the polarization of the frontier orbitals. The polarity of the frontier orbitals can often be deduced by combining the frontier orbitals of those conjugated systems with which they can be compared.

4 Electrocyclic reactions

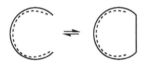

4.1 Introduction

Electrocyclic reactions are characterized by the creation of a ring from an open-chain conjugated system, with a σ-bond forming across the ends of the conjugated system, or, of course, the reverse of this reaction, the opening of a σ-bond with the creation of a longer conjugated system. Unfortunately, the word electrocyclic is sometimes used wrongly by the unwary when they really mean pericyclic. This mistake has come about because the word electrocyclic was introduced before there was any word to describe the whole family of pericyclic reactions, and some people have never caught up.

4.2 Neutral polyenes

Fig. 4.1 illustrates the first few members of the series of neutral polyenes: the equilibria between butadiene **4.1** and cyclobutene **4.2**, between hexadiene **4.3** and cyclohexadiene **4.4**, and between octatetraene **4.5** and cyclooctatriene **4.6**. There are of course heteroatom-containing analogues, with nitrogen or oxygen in the chain of atoms, and the systems can be decked out with substituents and other rings. To appreciate what the fundamental reaction is, it is only necessary to tease out the components—the longer conjugated system on one side of the equilibrium, and the σ-bond and the shorter conjugated system on

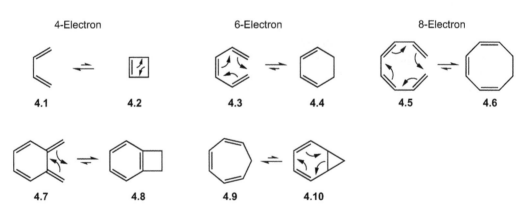

Fig. 4.1 4-, 6-, and 8-Electron electrocyclic reactions of conjugated hydrocarbons

the other. The strained ring of the cyclobutene **4.2** makes this reaction take place in the ring-opening sense, while the hexatriene and octatetraene reactions are ring closures, because in both cases a σ-bond effectively replaces a π-bond.

It is not difficult to find examples in which these reactions take place in the opposite direction. In contrast to the ring opening of the cyclobutene **4.2** and the ring closing of the hexatriene **4.3**, the butadiene component in o-quinodimethane **4.7** undergoes ring-closure to the cyclobutene **4.8**, and the cyclohexadiene component in the cycloheptatriene **4.10** undergoes ring-opening to give the hexatriene **4.9**.

Whether the reaction takes place in the opening or closing direction, we are immediately confronted with what looks like a paradox: these reactions take place when the total number of electrons is a $(4n + 2)$ number or a $(4n)$ number, in contrast to the pericyclic cycloadditions we saw in the previous chapter, which only take place when the total number of electrons is a $(4n + 2)$ number. We shall see shortly that it is the stereochemistry that changes, depending upon whether the total number of electrons is $(4n + 2)$ or $(4n)$.

4.3 Ionic conjugated systems

Fig. 4.2 illustrates the first few members of the series of equilibria of conjugated ions. In cations, they are the equilibria between the allyl **4.11** and the cyclopropyl cation **4.12**, the pentadienyl **4.13** and the cyclopentenyl cation **4.14**, and the heptatrienyl **4.15** and cycloheptadienyl cation **4.16**. In anions, they are between the allyl **4.17** and the cyclopropyl anion **4.18**, the pentadienyl **4.19** and the cyclopentenyl anion **4.20**, and the heptatrienyl **4.21** and cycloheptadienyl anion **4.22**. There are heteroatom-containing analogues, with nitrogen and oxygen lone pairs rather than a carbanion centre, and the systems can again have substituents and fused rings.

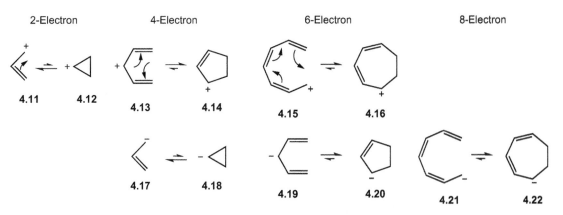

Fig. 4.2 2-, 4-, 6-, and 8-Electron electrocyclic reactions of conjugated hydrocarbon ions

These reactions are rarely seen in their unadorned state, and the direction in which they go is nearly always determined by the substituents or hetero-atoms stabilizing one side of the equilibrium or the other. But in all of them, the pathway between the open-chain and the cyclic isomers is the same, and the rules for stereochemistry are not dependent upon the direction in which the reaction actually takes place. They depend on the total number of electrons involved.

The equilibria between the allyl and the cyclopropyl cations and the corresponding equilibria of the anions, are not seen as such, but they are found in disguise. Thus the cationic reaction is seen when cyclopropyl halides are heated or treated with silver ions. The cyclopropyl cation itself is not an intermediate, because the ring opening (**4.23**, arrows) takes place at the same time that the halide ion leaves. The allyl cation **4.24** may possibly be an intermediate, but its lifetime must be very short, for the halide ion recaptures the cation on the same surface from which it has just left, without having time to drift round to the other side. In the reverse direction, the cyclization of an allyl cation is seen as a step in the Favorskii rearrangement, where the enolate **4.26** of an α-haloketone cyclizes to a cyclopropanone **4.27**, with the bromide leaving in concert with the electrocyclic reaction when the reaction is carried out in solvents of low polarity. Note that the cyclization is driven forward because the oxy-anion of the enolate stabilizes the cyclopropyl cation in the intermediate—a carbonyl group, in other words, is a highly stabilized carbocation, as we have seen earlier. The final step of the Favorskii reaction, the ring opening to give the ester **4.28**, is not, of course, a pericyclic reaction.

A halide ion departing initiates the electrocyclic opening of a cyclopropyl cation, without the cation itself being an intermediate.

4.23   4.24   4.25

A halide ion departing initiates the electrocyclic closing of an allyl cation, without the cation itself being an intermediate.

4.26   4.27   4.28

The equilibrium between an allyl anion and a cyclopropyl anion is represented by the reversible opening of epoxides and aziridines, where the lone pair on the oxygen or nitrogen takes up the role of the carbanion. The opening needs some heat—and it only occurs easily when the carbon atoms have anion-stabilizing groups. It is detected by the subsequent 1,3-dipolar cycloadditions that the carbonyl ylids or the azomethine ylids, undergo. Thus, heating tetracyanoethylene oxide **4.29** gives a carbonyl ylid **4.30**, which can be trapped as a tetrahydrofuran **4.31** by alkenes.

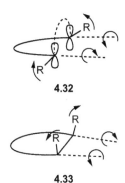

The lone pair on oxygen makes an epoxide isoelectronic with a cyclopropyl anion.

Some flexibility of mind is required to recognize that reactions like these are electrocyclic, but they certainly are electrocyclic, as shown by their stereochemistry, their single most remarkable feature.

4.4 **Stereochemistry**

There are two possible stereochemistries for the ring-closing and ring-opening reactions. They are called **disrotatory** and **conrotatory**, and are illustrated for the general cases in Fig. 4.3. Looking at the ring-closing disrotatory reaction **4.32**, the two outer substituents R move upwards, so that the top lobes of the p-orbitals turn towards each other to form the new σ-bond. The word disrotatory reflects the fact that the rotation about the terminal double bonds is taking place clockwise at one end but anticlockwise at the other. In the corresponding ring-opening **4.33**, there is similarly a clockwise and anticlockwise rotation as the

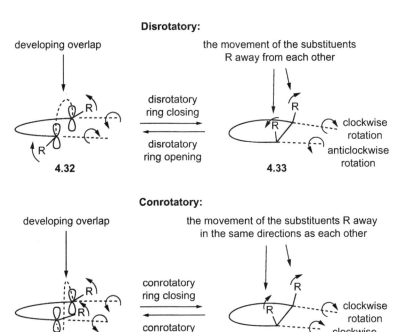

Fig. 4.3 Definitions and features of disrotatory and conrotatory reactions

σ-bond breaks, and the two substituents that start off *cis* to each other move apart to become the outer substituents in the open-chain conjugated system. There is an equally probable disrotatory ring closure, in which both R groups fall, with the lower lobes of the p-orbitals forming the new σ-bond, and there is a possible alternative disrotatory ring opening, in which both R groups move towards each other, although whether this happens depends upon the size of the R groups and the extent to which they meet steric hindrance by moving inwards.

In contrast, in conrotatory ring-closing **4.34 → 4.35**, one of the outer substituents and one of the inner substituents, both labelled R, rise to become *cis*, so that the bottom lobe of the p-orbital at one end forms a σ-bond by overlap with the top lobe of the p-orbital at the other end. The rotations are now in the same sense, either both clockwise or both anticlockwise. It follows that the outer substituents become *trans* to each other on cyclization. In the ring-opening, **4.35 → 4.34**, the two substituents that are *cis* to each other move in the same direction, one to an outer position and the other to an inner position by clockwise rotations, as drawn here. Alternatively, of course, they could both move by anticlockwise rotations. The rules for which stereochemistry is followed by which system are these:

The Woodward–Hoffmann rules for electrocyclic reactions:

	Thermal	Photo
Two Electrons:		
allyl cation-cyclopropyl cation	disrotatory	conrotatory
Four Electrons:		
allyl anion-cyclopropyl anion	conrotatory	disrotatory
butadiene-cyclobutene	conrotatory	disrotatory
pentadienyl cation-cyclopentenyl cation	conrotatory	disrotatory
Six Electrons:		
pentadienyl anion-cyclopentenyl anion	disrotatory	conrotatory
hexatriene-cyclohexadiene	disrotatory	conrotatory
heptatrienyl cation-cycloheptadienyl cation	disrotatory	conrotatory
Eight Electrons:		
heptatrienyl anion-cycloheptadienyl anion	conrotatory	disrotatory
octatetraene-cyclooctatriene	conrotatory	disrotatory
nonatetraenyl cation-cyclononatrienyl cation	conrotatory	disrotatory

These rules look difficult to absorb all at once, but a simplification makes them easy to learn—all thermal reactions involving a total number of electrons that can be expressed in the form $(4n + 2)$ are disrotatory, and the others, in which the total number of electrons can be expressed in the form $4n$, are conrotatory. The rules for the corresponding reactions in the first electronically excited state are simply the opposite.

The only electrons that count in the open-chain component are those in the continuous linear array of p-orbitals. Substituents, whether at the ends of the conjugated system or anywhere else along its length, do not affect the stereochemistry, although they often affect the rate or the position of equilibrium, as we saw in the effect of the oxyanion substituent in the cyclization **4.26 → 4.27**.

4.5 The Woodward–Hoffmann rules applied to thermal electrocyclic reactions

As a first example of an electrocyclic reaction illustrating stereochemistry, let us take a pair of conrotatory cyclobutene openings, showing that the reactions are stereospecific.

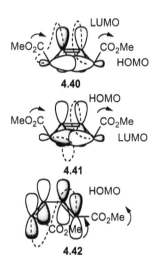

The fact that the reactions take place in the direction of ring-opening is determined by thermodynamics, but the stereochemistry is most certainly not, for the cyclobutene **4.36** gives the thermodynamically higher-energy product **4.37** with one of the double bonds *cis*. Thermodynamics affects the stereochemistry only with the opening of the cyclobutene **4.38**, which shows a preference for one of the conrotatory modes, that giving the *trans,trans* diene **4.39**, where the rules could have led to the *cis,cis* diene equally well. This type of selectivity is called **torqueoselectivity**.

As with cycloadditions, there are three explanations for why these reactions take the stereochemical path they do. The aromatic transition structure is accommodated by the idea of a Möbius conjugated arrangement in the orbitals for cyclobutene opening. Although the idea of a Möbius conjugated system is simple, it is not obvious. Frontier orbital theory is obliged, somewhat artificially, to treat the σ-bond and the π-bond in the ring opening as separate entities. It does not matter which of them one takes for the HOMO and which for the LUMO—they both work **4.40** and **4.41**. There is a bigger problem with ring-closure, for it is not possible to have a HOMO and LUMO when the diene is treated as only one component, and frontier orbital theory relies upon there being a pair of interacting orbitals. Although the choice of the HOMO appears arbitrary, it was the observation that the signs of its coefficients **4.42** matched the conrotatory stereochemistry for the cyclobutene-butadiene reaction (and that they also matched for the disrotatory cyclohexadiene-hexatriene reaction) that first led Woodward to turn to molecular orbital theory for an explanation of the starkly contrasting stereochemistry of these reactions. The outcome, independently developed by Hoffmann and by Abrahamson and Longuet-Higgins, was the correlation diagram that we have already seen applied to cycloaddition reactions—correlation diagrams were in fact first applied to these electrocyclic reactions, and they remain the most satisfying explanation.

We shall look only at the orbital correlation diagram for cyclobutene opening (Fig. 4.4). Conrotatory cyclobutene opening preserves an axis of symmetry running through the centres of both the σ- and the π-bond. Taking the π-bonding orbital **4.43** as an example, note how the symmetry classification works: the filled lobe is reflected through the axis (the dashed line passing through the dot in the middle of the bond) by an unfilled lobe, making it antisymmetric, whereas it would be symmetric with respect to the plane of symmetry maintained in a disrotatory opening. Fig. 4.4 shows that the filled orbitals of the ground state of the starting material correlate with the filled orbitals of the ground state of the product. If you try a correlation diagram with a plane of symmetry, you will find that the orbitals of the ground state in the starting material do not correlate with the orbitals of the ground state in the product. They correlate instead with the orbitals of a doubly-excited state, with the result that the disrotatory opening meets the same high symmetry-imposed barrier as a [2 + 2] cycloaddition.

unfilled lobe

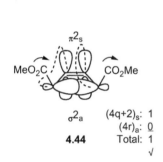

filled lobe

4.43

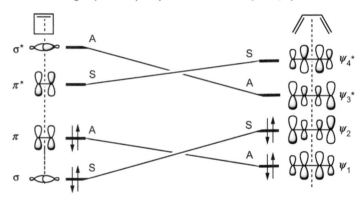

Fig. 4.4 Correlation diagram for the conrotatory opening of a cyclobutene

This is the last correlation diagram we shall see—from now on we shall accept that they explain pericyclic stereochemistry convincingly, and shall simply use the Woodward–Hoffmann rules that developed from them. Cyclobutene opening uses four electrons, a 4n number, which means that one antarafacial component is needed. Using the interconversion of the cyclobutene **4.36** and the butadiene **4.37** as an example, the forward reaction is essentially treated as the addition of the σ-bond to the π-bond, in which either the σ- or the π-bond can take up the role of antarafacial component, **4.44** or **4.45**, and still be the same reaction. It is called $[_\sigma2_a + _\pi2_s]$ in one and $[_\sigma2_s + _\pi2_a]$ in the other, emphasizing the fact that these descriptions apply to the drawings, and do not uniquely describe the reaction. In the backwards direction **4.46**, all that is needed is to treat the diene as the one antarafacial component, with the top lobe on the right tilting over as the bottom lobe on the left tilts up to meet it in a $[_\pi4_a]$ reaction. Notice how the dashed lines correspond to the developing overlap that tilts the substituents down and to the right in **4.44** and **4.45**, and up and to the left in **4.46**, leading to the correct isomer in each direction by a conrotatory movement.

$\pi2_s$

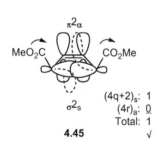

MeO$_2$C   CO$_2$Me

$\sigma2_a$ (4q+2)$_s$: 1
                  (4r)$_a$: 0
4.44   Total: 1
                        √

$\pi2_a$

MeO$_2$C   CO$_2$Me

$\sigma2_s$ (4q+2)$_s$: 1
                  (4r)$_a$: 0
                  Total: 1
4.45   √

$\pi4_a$

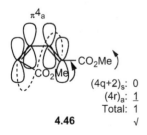

CO$_2$Me
CO$_2$Me

                  (4q+2)$_s$: 0
                  (4r)$_a$: 1
                  Total: 1
4.46   √

We see the same pattern, but with opposite stereochemistry, in the disrotatory reactions of the hexatrienes **4.47** and **4.49**. The direction the reaction takes is determined by thermodynamics, ring-closing in this case, but the stereochemistry is not, since the *cis* disubstituted cyclohexadiene **4.48** is higher in energy than its *trans* isomer. If you try the frontier orbitals or a correlation diagram with a plane of symmetry, you will find that they explain this stereochemistry.

The hexatriene reaction is slow, because the unstrained transition structure, a graceful spiral bringing the p-orbitals easily within bonding distance, corresponds to the forbidden $[_\pi 6_a]$ conrotatory process.

With two more electrons in the conjugated system, making eight in all, the octatetraenes **4.51** and **4.54** show conrotatory closure giving the cyclo-octatrienes **4.52** and **4.55**. However, the reaction can only just be stopped at this stage, for the products undergo a second electrocyclic reaction giving the bicyclic dienes **4.53** and **4.56** as a result of the allowed disrotatory reaction of the all-*cis* hexatriene units.

Instead, the transition structure must involve the lobes on the top surface tilting towards one another, in order to achieve an allowed $[_\pi 6_s]$ disrotatory process.

The reaction of the octatetraene is faster, because the easily achieved spiral transition structure corresponds to the allowed $[_\pi 8_a]$ conrotatory process.

In this example the ring system is compatible with the allowed stereochemistry; the disrotatory equilibrium between **4.55** and **4.56** has no problems. On the other hand, rings can constrain or even prevent allowed electrocyclic reactions. In the cyclobutene **4.57**, for example, the ring fusion is the lower-energy *cis* arrangement, and the double bonds within the rings are all *cis*. As a result, the conrotatory opening of the cyclobutene **4.57** gives a diene **4.58** with a strained *trans* double bond within the eight-membered ring—which is thermo-dynamically unfavourable. The reaction therefore needs a high temperature, but it can be detected by the deuterium-labelling experiment, which gives the isomer **4.59**.

The difference in activation energy between the allowed and the forbidden reactions is considerable. Wherever the activation enthalpy has been estimated for both reactions, it has come to more than 45 kJ mol⁻¹ in favour of the allowed pathway. All the reactions illustrated above show no trace of the forbidden isomer. In the most delicate measurement yet performed, the conrotatory opening of *cis*-3,4-dimethylcyclobutene gave less than 0.005% of *trans,trans*-hexa-2,4-diene!

With a smaller ring in the bicycloheptene **4.60**, a conrotatory reaction is virtually impossible, since it would put a *trans* double bond into a seven-membered ring. No visible reaction occurs until the forbidden disrotatory reaction **4.62** gives the *cis,cis*-diene **4.61** at 400°, and even then only in low yield.

A symmetry-forbidden reaction can take place, but it requires much more energy to drive it:

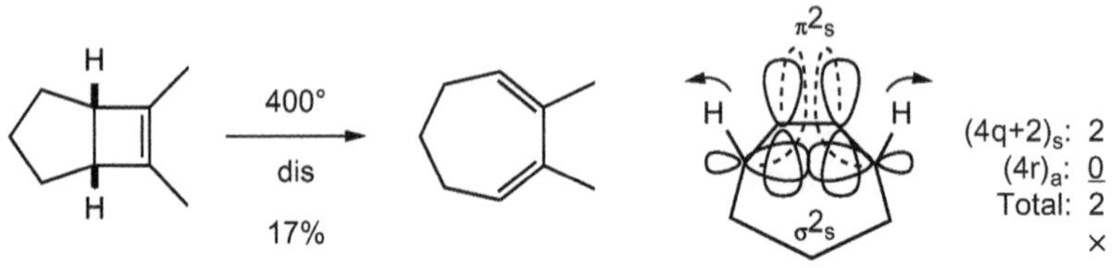

$\pi^2{}_s$

$(4q+2)_s$: 2
$(4r)_a$: 0
Total: 2
×

| 4.60 | 400°
dis
17% | 4.61 | 4.62 |

The electrocyclic reactions of ions equally follow the Woodward–Hoffmann rules. The opening of the cyclopropyl halide **4.23** must have been disrotatory, since it produces the cyclic allyl cation **4.24**. One might argue that it has no alternative, but the reaction proves to be disrotatory in open-chain systems too, with the cyclopropyl halides **4.63**–**4.65** giving the allyl cations **4.66**–**4.68**, as shown by their distinct and diagnostic ¹H NMR spectra. The allyl cations have three possible configurations— W-shaped **4.66**, sickle shaped **4.67**, and U-shaped **4.68**, which largely retain their configuration at low temperature because of the bonding overlap between C-1,

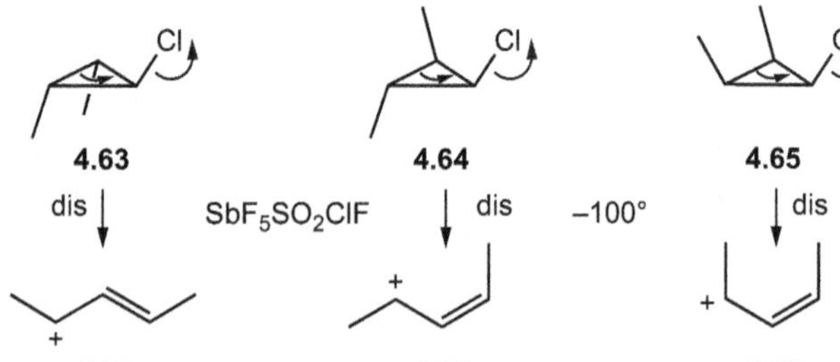

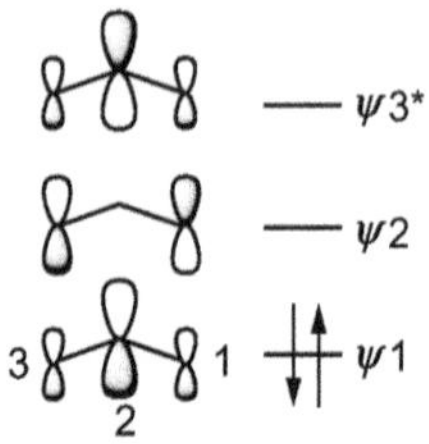

C-2, and C-3 (with the same orbitals **4.69** as the anion in Chapter 2 on p. 21 and the radical p. 29). The U-shaped ion has a half life of 10 min at −10° and the sickle-shaped ion a half life of 10 min at 35°. These reactions demonstrate torqueoselectivity—the direction of disrotatory opening is determined by which side of the ring the halide leaves from, as shown by the contrast between the products from the halides **4.63** and **4.65**. The former opens with the two methyl groups moving outwards, in what must be the thermodynamically preferred direction, but the latter evidently eschews thermodynamics, and opens to give the U-shaped ion. Their giving different products is further evidence that a free cyclopropyl cation is not involved, but the strict disrotatory pathway shows that the reactions must nevertheless be pericyclic in nature. It appears that thermodynamics is less important in determining the reaction pathway than the strong torqueoselectivity combined with the demands of orbital symmetry.

A simple explanation for the torqueoselectivity is to think of the bond between C-2 and C-3 in the chloride **4.65** bending downwards, moving its electrons in behind C-1, from which the chloride ion is leaving, as though they were carrying out an S_N2 reaction on it. It follows that, if the bond is moving downwards the substituents must be moving towards each other to give the U-shaped allyl cation **4.68**. To see how this conforms to the Woodward–Hoffmann rule, one simply counts the breaking C–Cl bond, with the electron pair moving towards the chlorine, as an empty p-orbital **4.70**. The disrotatory reaction is then the allowed combination $[_\sigma2_s + _\omega0_s]$, suprafacial on both components. The small sum that has accompanied all the drawings so far will be left out from now on—readers can check for themselves that they all fit the rule.

The same reaction in the opposite direction—the ring closure of an allyl cation—is also known. The remarkable formation of the thermodynamically less favoured *cis*-di-t-butylcyclopropanone **4.72** from the zwitterion **4.71**, which probably has the W configuration, is evidence of its being disrotatory.

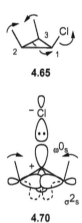

4.65

4.70

4.71   **4.72**

4.73   **4.74**   **4.75**

This result contrasts nicely with a similar system with two more electrons. The heterocycle **4.73** loses nitrogen to give the thiocarbonyl ylid **4.74**, which has the sickle-shaped configuration, because the 1,3-dipolar cycloreversion is the reverse of a suprafacial addition. This ylid, with four electrons in the conjugated system, undergoes conrotatory cyclization to give the di-t-butylthietan **4.75**, which again has the two large groups *cis* to each other. In the Woodward–Hoffmann scheme **4.76** the conrotatory closure will be a $[_\pi4_a]$ reaction.

In the ring-opening direction with four electrons, we saw a pair of azomethine ylids **2.98** and **2.100** in Chapter 2 on p. 22 redrawn here as **4.78** and **4.80**. These ylids had been produced by heating the aziridines **4.77** and **4.79**, respectively, in a conrotatory reaction, which is now seen to be stereospecific, since both isomers are available. The proof of the stereochemistry of the intermediates **4.78** and **4.80** comes from the subsequent, suprafacial cycloaddition, drawn on p. 22, that traps

4.76

the ylids before they lose their W- and sickle-shaped configurations. The alternative conrotatory opening of the aziridine **4.77** is not seen—it would be just as allowed by symmetry, but this time there is no special reason to form the U-shaped ylid in preference to the thermodynamically preferred W-shaped ylid **4.78**.

4.77   **4.78**   **4.79**   **4.80**

4.81

The conrotatory stereochemistry fits the Woodward–Hoffmann rule, illustrated in **4.81** as $[_\sigma 2_s + _\omega 2_a]$ for the conversion of the *cis* aziridine **4.79** into the sickle-shaped ylid **4.80**. The dashed lines could equally have been drawn to make it $[_\sigma 2_a + _\omega 2_s]$.

A pentadienyl cation has the same number of π-electrons as the allyl anion, and its electrocyclic reactions will be conrotatory. It has been shown to be fully stereospecific, with the stereoisomeric pentadienyl cations **4.82** and **4.84** giving the stereoisomeric cyclopentenyl cations **4.83** and **4.85** in conrotatory reactions, followed in their 1H NMR spectra. In terms of the Woodward–Hoffmann rule, it can be drawn **4.86** as an allowed $[_\pi 4_a]$ process.

4.82   **4.83**   **4.84**   **4.85**

4.86

Perhaps the most remarkable feature of this reaction is that a bond has formed between C-1 and C-5, both of which are positively charged. Any attempt to think of this reaction as the combination of a nucleophilic and an electrophilic carbon would not make proper sense, yet the reaction occurs easily. Pericyclic reactions really are a distinctly different class of reactions from ionic and radical reactions. Since this reaction is also 5-*endo-trig* at both ends, it would appear to be also deeply forbidden by Baldwin's rules—which evidently do not apply with any great force to electrocyclic reactions.

The synthetically most useful reaction of this type is the Nazarov cyclization, in which a cross-conjugated dienone like **4.87** forms the cyclopentenones **4.90** and **4.91** when treated with acid. Protonation of the ketone makes it more of a carbocation, and hence the conjugated system more of a pentadienyl cation **4.88**. The conrotatory cyclization takes place to give the cyclopentenyl cation **4.89**, which loses a proton in either of two directions before picking up a proton to give the ketones **4.90** and **4.91**. The relative stereochemistry at C-1 and C-5 has the two hydrogen atoms *trans*, proving that the cyclization has been conrotatory, but this is only confirmed in the minor product **4.91**. The major product, with the proton lost mainly from C-1, no longer gives the stereochemical information.

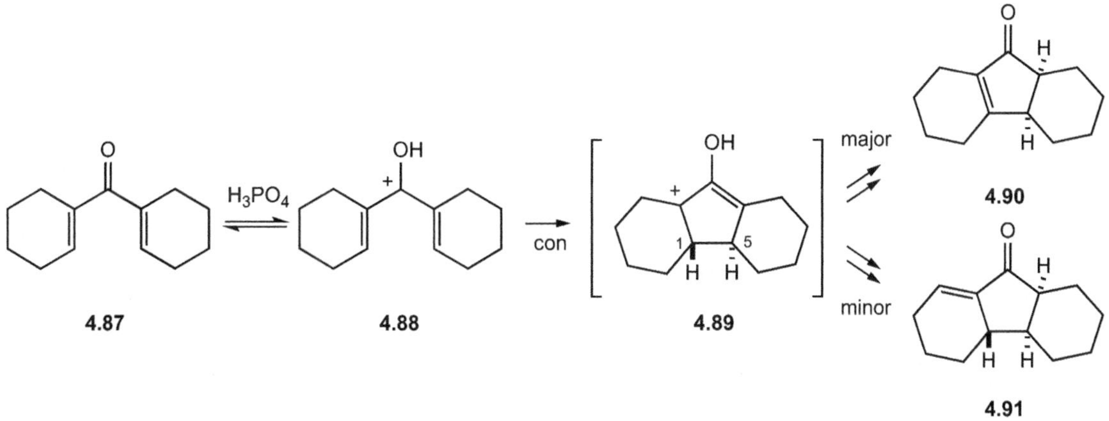

The solution to this problem is achieved by using a better electrofugal group than a proton. The silylated dienone **4.92** undergoes the same reaction, catalysed by a Lewis acid, but this time the intermediate cation exclusively loses the silyl group to give the product **4.91**.

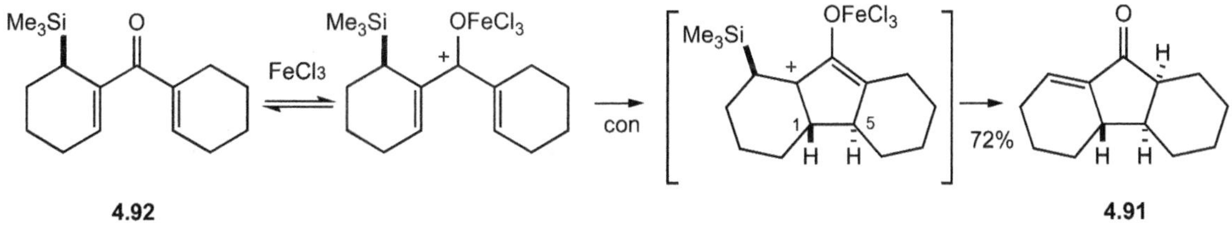

A remarkable feature of this reaction is that the side of the ring on which the silyl group is placed (the upper surface here) determines which of the two equally allowed conrotatory ring closures actually takes place, in other words its torqueoselectivity. When the ketone **4.92** is enantiomerically enriched, the ketone **4.91** is formed enantiomerically enriched to the same extent, and in the sense which corresponds to the specific anticlockwise conrotatory movement **4.93**, with the lower lobe on C-1 turning upwards, presumably because it is *anti* to the silyl group, which is a large donor substituent.

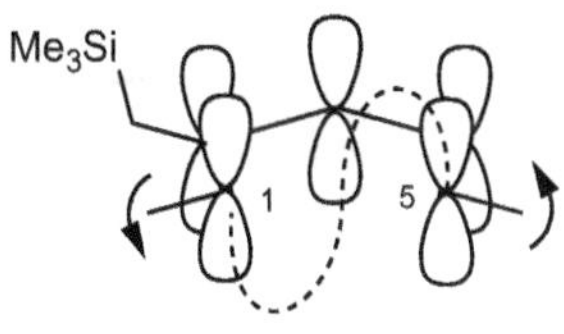

Pentadienyl anions ought to close in a disrotatory fashion, since they can be drawn as an allowed [$_\pi 6_s$] process **4.94**. The anion **4.96** generated from hydrobenzamide **4.95** gives amarine **4.97**, in which the two phenyl groups are *cis*. This reaction, first observed in 1844, is probably the oldest known pericyclic reaction (the composer Borodin published a paper about it). The fact that it gives the thermodynamically less favoured product, with the two large substituents *cis*, is evidence that it must have been disrotatory, although the full stereospecificity has not been proved. Further circumstantial evidence for rule-obeying behaviour in these reactions comes from the observation that irradiation of the anion **4.96** gives the *trans* isomer of **4.97**.

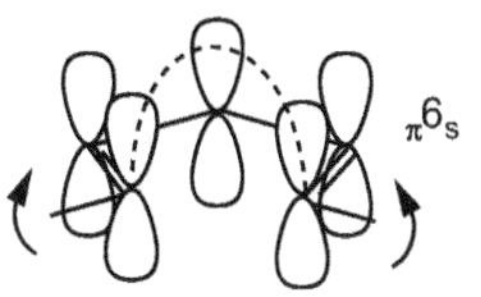

4.93

4.94

4.95   **4.96**   **4.97**

4.98

A disrotatory ring-opening of a cyclopentenyl anion has also been seen, when the relief of strain in a cyclopropane makes it thermodynamically favourable– the cyclopentenyl anion **4.100** opens to the pentadienyl anion **4.101**. This reaction had no option but to be disrotatory in the sense in which the two hydrogen atoms move outwards **4.98**, since a *trans* double bond would be impossible in the six-membered ring.

4.99   **4.100**   **4.101**

4.6 Photochemical electrocyclic reactions

As with cycloadditions, photochemical reactions that look pericyclic may not always be pericyclic, but several do seem to obey the rules. The four-electron photochemical reaction **4.102**→**4.103** is disrotatory and the six-electron reaction **4.104**→**4.105** is conrotatory. The reactions are counter-thermodynamic, with the products the less stable stereoisomer, in both cases.

4.102   **4.103**   **4.104**   **4.105**

4.106   **4.107**

When the products of these two reactions are heated, they give back isomers of the starting materials, the *trans,cis* diene **4.106** in place of **4.102**, and the *trans*-dimethylcyclohexadiene **4.107** in place of **4.104**.

In more constrained systems, one of the consequences of the photochemical and thermal rules' being opposite is that a photochemical reaction can easily lead to a product thermodynamically much higher in energy than the starting material, but nevertheless thermally stable. The photochemical reactions **4.108**→**4.109** and **4.110**→**4.111** illustrate this point. The bicyclic ketone **4.109** is not only strained, but has also lost both the partial aromaticity of the tropone **4.108** and the conjugation between the methoxy group and the ketone. The dihydroanthracene **4.111** has lost the aromaticity of both benzene rings present in *cis*-stilbene **4.110**. The bicyclic ketone **4.109** and the dihydroanthracene **4.111** meet high symmetry-imposed kinetic barriers in trying to return thermally to their precursors, and the rule-obeying processes are impossible because they would produce *trans* double bonds in the rings of the starting materials.

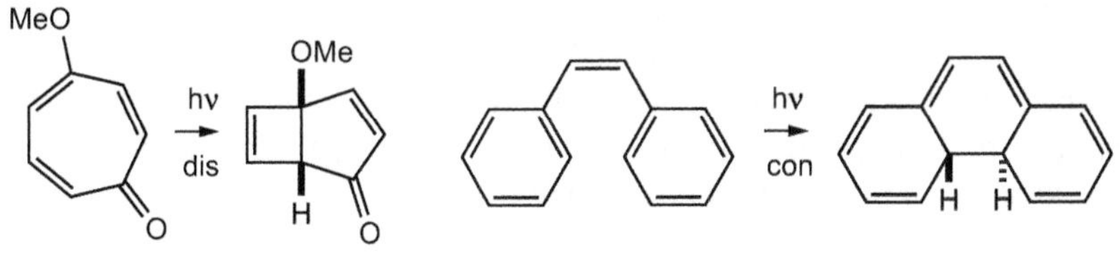

4.108 **4.109** **4.110** **4.111**

The restraints imposed by rings are especially well demonstrated by some of the oldest known electrocyclic reactions (Fig. 4.5), puzzling in their day, but now beautiful examples illustrating the rules of electrocyclic reactions. The full structure is ergosterol **4.112**, abbreviated to **4.113** in Fig. 4.5.

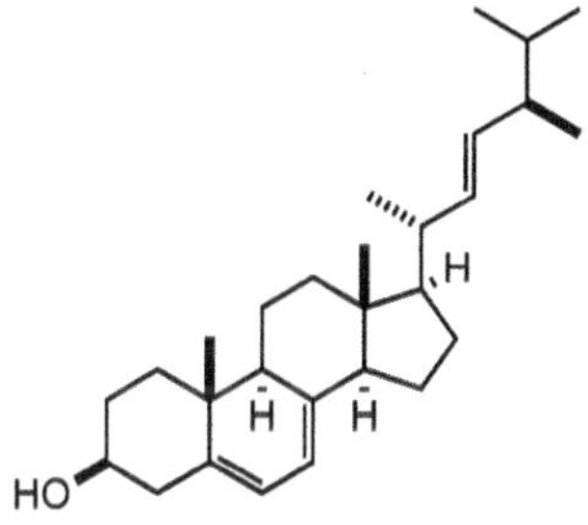

4.112

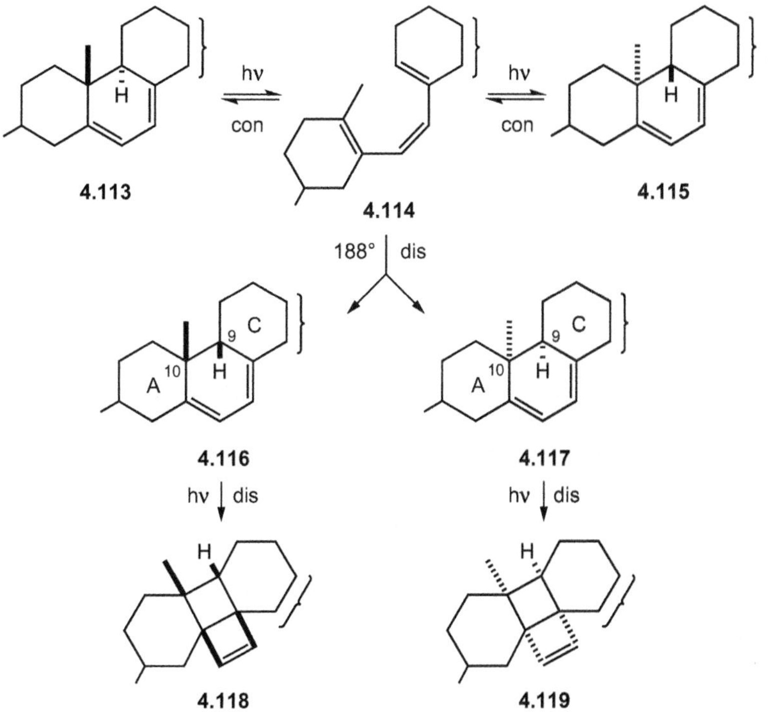

4.113 **4.114** **4.115** **4.116** **4.117** **4.118** **4.119**

Fig. 4.5 Photochemical and thermal electrocyclic reactions of ergosterol

We saw the product **4.114** from the photochemical conrotatory opening of ergosterol as **1.21** in Chapter 1 on p. 5, and we shall see its analogue **5.9** in Chapter 5 on p. 84. It is in photochemical equilibrium with its starting material **4.113**, but it is also in photochemical equilibrium with the product **4.115** of the alternative conrotatory ring closure. Thermally, however, it gives a mixture of the other two cyclohexadienes **4.116** and **4.117** in disrotatory cyclizations. When these compounds, with the methyl and hydrogen groups on C-10 and C-9 *cis* to each other, are irradiated, the allowed conrotatory opening would put a *trans* double bond either into ring A or into ring C. To avoid this calamity, a different allowed photochemical reaction takes place, namely the disrotatory closures to the cyclobutenes **4.118** and **4.119**. These compounds in turn are thermally comparatively stable, because the allowed conrotatory thermal ring-openings would place a *trans* double bond into ring B.

4.7 **Further reading**

All the less specialized books mentioned in the last chapter have sections on electrocyclic reactions. A specialized treatise is:

E. N. Marvell, *Thermal Electrocyclic Reactions*, Academic Press, New York, 1980.

Cyclobutene ring-opening:

T. Durst and L. Breau, in *COS*, Ch. 6.1;

Cyclohexadiene reactions:

W. H. Okamura and A. R. De Lera, in *COS*, Ch. 6.2;

Nazarov:

S. E. Denmark, in *COS*, Ch. 6.3.

4.8 **Problems**

4.1 Draw the orbitals for the hexatriene-cyclohexadiene reaction **4.47** → **4.48** and its reverse, in the style **4.44**, **4.45**, and **4.46** used for the corresponding cyclobutene-butadiene reaction, identify the developing overlap, and hence show that the symmetry-allowed reaction is disrotatory in both directions.

4.2 Explain how the stereoisomeric butadienes **4.120** and **4.121** equilibrate without forming any of the isomer **4.122**.

4.120   **4.121**   **4.122**

4.3 Explain how the enantiomerically enriched *trans*-2,3-di-t-butylcyclopropanone **4.123** racemizes at 80°, but does not form even a trace of the *cis* isomer **4.125**.

4.123   **4.124**   **4.125**

4.4 Predict the stereochemistry of the products of these reactions:

(a)

(b)

(c)

(d)

(e)

4.5 Account for the relative stereochemistry in the product of this Staudinger reaction:

4.6 The following reactions take place with combinations of electrocyclic reactions, cycloadditions and retro-cycloadditions, in any order. Identify the steps, 2 in each of (a), (b), and (c), and five in (d):

(a)

(b)

(c)

(d)

4.9 Summary

- Electrocyclic reactions are unimolecular; they are either cyclizations, in which a σ-bond is formed across the ends of a π-conjugated system, or the reverse of such a reaction.

- Electrocyclic reactions are also seen in cations, in anions, and in conjugated systems using lone pairs of electrons isoelectronic with anions.

- Some electrocyclic reactions take place in concert with the ionization of a nucleofugal group, before the free cation is formed.

- Electrocyclic reactions take place in the direction determined by thermodynamics, with ring strain often favouring ring-opening.

- The same rule governs electrocyclic reactions as cycloadditions: a ground-state pericyclic change is symmetry-allowed when the total number of $(4q + 2)_s$ and $(4r)_a$ components is odd.

- When the total number of electrons can be described in the form $(4n + 2)$, a plane of symmetry is maintained in the orbitals undergoing a change, and the stereochemical outcome is described as disrotatory.

- When the total number of electrons can be described in the form $(4n)$, an axis of symmetry is maintained in the orbitals undergoing change, and the stereochemical outcome is described as conrotatory.

- The same explanations for the stereochemical outcome work as they did for cycloadditions: the aromatic transition structure (with Möbius aromaticity in those reactions mobilizing 4n electrons), the frontier orbital picture (artificially separating the σ-bond and the π-system), and, most satisfyingly, correlation diagrams.

- In both disrotatory and conrotatory reactions, there are two directions of twist allowed by the rules. Which of these directions is followed is described as torqueoselectivity. Torqueoselectivity may be governed by steric effects, with large groups moving away from each other when the rules allow it, or by the constraints of a ring system making one of the pathways impossible. In some cationic reactions torqueoselectivity is determined by the side from which the nucleofugal group departs.

- Many electrocyclic reactions are *endo*-trig in three-, four-, and five-membered rings; Baldwin's rules do not apply to electrocyclic reactions.

- Photochemical electrocyclic reactions follow the opposite rules: conrotatory with (4n + 2) electrons and disrotatory with (4n).

Sigmatropic rearrangements

5.1 Introduction

Sigmatropic rearrangements are characterized by the movement of a σ-bond from one position to another, with a concomitant movement of the conjugated systems to accommodate the new bond and fill in the vacancy left behind.

There are two main families. In one, a single group or atom moves from one end of a conjugated system n atoms along to the other end. These are labelled as [1,n] shifts. The label [1,n] comes from the symbols in bold in the drawing. The migrating bond moves from C-1 to C-**n** along the conjugated system—and the **1** comes from the fact that the bond remains attached to the same atom R, without migrating along any conjugated system within the R group; it does not come from the fact that the bond starts off from C-1, which is C-1 by definition.

In the other family, the bond that moves is internal within two conjugated systems, and it can move **m** atoms along one conjugated system and the same or a different number **n** atoms along the other. These will then be labelled as [m,n] shifts.

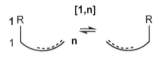

5.2 [1,n] Rearrangements—suprafacial and antarafacial

Pericyclic [1,n] sigmatropic rearrangements are most frequently shifts of hydrogen atoms, with examples known for $n = 2, 3, 4, 5, 6, 7$, and longer, some of which have a total of $(4n + 2)$ electrons involved and some $(4n)$. As with electrocyclic reactions, the stereochemistry changes from one series to the other. If the hydrogen leaves one surface of the conjugated system, and arrives at the other end on the same surface, the reaction is described as **suprafacial**, and this is the allowed pathway when the total number of electrons is $(4n + 2)$. Alternatively, if it leaves one surface and arrives on the opposite surface, it is called **antarafacial**, and this is the allowed pathway when the total number of electrons is $(4n)$. This is apt to be confusing, because these reactions may or may not be described in Woodward–Hoffmann terms as having suprafacial or antarafacial *components*, so the words do not mean exactly the same as they did before. The connection, however, will become clear with use, and Fig. 5.1 illustrates two straightforward examples.

Take the most common of all these reactions, the [1,5] hydrogen shift, which illustrates a suprafacial shift, in which the hydrogen leaves the upper surface at

C-1, and arrives on the upper surface at C-5. This can be drawn **5.1** as a $[_\sigma 2_s + _\pi 4_s]$ process, rather as though it was a cycloaddition of the σ-bond to the diene, and the connection between the two usages of the word suprafacial is evident. On the other hand, it could equally well have been drawn **5.2** as a $[_\sigma 2_a + _\pi 4_a]$ process; yet it is still the same reaction. It is *structurally* a suprafacial shift, but the developing overlap that happens to be *illustrated* in **5.2** has been described, perfectly correctly, as antarafacial on both components. In both cases, of course, the reaction is allowed by the Woodward–Hoffmann rule for thermal reactions.

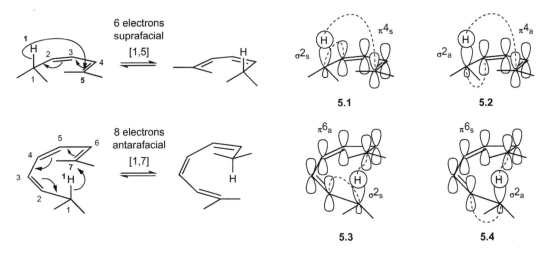

Fig. 5.1 Symmetry-allowed [1,5] suprafacial and [1,7] antarafacial shifts of hydrogen atoms

In the [1,7] hydrogen shift, the allowed pathway is an antarafacial shift, in which the hydrogen atom leaves the upper surface at C-1, and arrives on the lower surface at C-7. This can be drawn **5.3** as a $[_\sigma 2_s + _\pi 6_a]$ process or **5.4** as a $[_\sigma 2_a + _\pi 6_s]$ process. This time it is *structurally* an antarafacial shift, but the developing overlap that happens to be *illustrated* can be described with one suprafacial and one antarafacial component either way round. It is helpful to draw as many suprafacial components as possible, that is, preferring **5.1** to **5.2**, since the structurally suprafacial reaction is then also described with suprafacial overlap developing. Similarly it is helpful to draw **5.3** rather than **5.4**, since that makes the antarafacial component the triene system, from one side of which to the other the antarafacial shift of the hydrogen is taking place.

The stereochemistry has been proved to follow the allowed pathways. Heating the diene **5.5** induces a suprafacial hydrogen shift to give the diene **5.6**, a suprafacial deuterium shift then converts this diene into the diene **5.7**, and another suprafacial deuterium shift converts it into a fourth isomer **5.8**. The major components at equilibrium are the isomers **5.6**, with an *E* trisubstituted double bond and *S* at the stereogenic centre, and **5.8**, with a *Z* trisubstituted double bond and *R* at the stereogenic centre. Neither of the other possible isomers, *E,R* or *Z,S*, is evident, showing that no [1,5] antarafacial shifts had occurred.

5.5   **5.6**

5.8   **5.7**

The antarafacial nature of the [1,7] shift was first inferred from the observation that they were known only in open-chain systems like that shown in Chapter 1 as **1.20→1.21**. More recently it has been proved by equilibrating the triene **5.10** with the two products of [1,7] shifts, **5.9** and **5.11**, the former from antarafacial shift of the hydrogen atom and the latter from antarafacial shift of the deuterium. There is no trace of either of the alternative isomers in which hydrogen has shifted to the top surface or deuterium to the bottom of C-7 (C-10 in steroid numbering).

5.9   **5.10**   **5.11**

One of the most common [1,5] hydrogen shifts takes place in cyclopentadienes, where it is constrained to be suprafacial. Because the atoms C-1 and C-5 are held close together by a σ-bond, the reactions **5.12→5.13→5.14** take place even at room temperature. It is important to recognize that mechanistically this is a [1,5] and not a [1,2] shift. The hydrogen atom does indeed move to the adjacent atom, so that it is structurally a 1,2 shift, but the rearrangement is possible only because the migrating σ-bond is conjugated through the π-system to C-5. The presence of the σ-bond between C-1 and C-5 is not essential—it merely serves to bring the reaction centres close together, speeding up the reaction, but not changing its fundamental nature.

5.12   **5.13**   **5.14**

The forbiddenness of a suprafacial [1,7] shift is shown by the absence of the same type of reaction in cycloheptatrienes **5.15**. Heating these compounds, when a substituent is present to reveal the presence of a reaction, interconverts all the mono-substituted isomers, just as it did with the cyclopentadienes, but the order in which the isomers appear shows that the pathway is a series of allowed suprafacial [1,5] shifts **5.15**→**5.16**→**5.17**→**5.18**. Since the hydrogen atom is not held close to C-5, these reactions are noticeably slower than the [1,5] shifts in cyclopentadienes. Photochemically, [1,5] hydrogen shifts occur only in open-chain systems, where they can be antarafacial. Photochemical [1,7] shifts can be suprafacial, and this is the preferred path in cycloheptatrienes, with the isomer **5.15** giving **5.17** as the first product in the series, followed by **5.18** and **5.16**, in contrast to the [1,5] suprafacial thermal reactions.

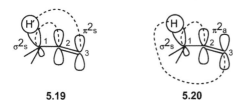

     5.15   **5.16**   **5.17**   **5.18**

Photochemically, a [1,3] suprafacial shift of hydrogen **5.19** is allowed, and has been observed with several 1,3-dienes rearranging to the unconjugated 1,4-dienes on irradiation, but the possibility of finding a thermal [1,3] shift of hydrogen is remote. It is clear that the structurally plausible suprafacial shift **5.19** is forbidden for a thermal reaction—it is $[_\sigma 2_s + _\pi 2_s]$. In principle, an antarafacial shift is thermally allowed **5.20**, but it is exceedingly unlikely ever to be found, because of the difficulty of maintaining π-overlap between C-1 and C-2 at the same time as the hydrogen atom reaches round to bond to the lower surface of C-3.

                5.19   **5.20**

With elements other than hydrogen, two new possibilities arise. First, metals and halogens show a capacity to undergo sigmatropic rearrangements, but the mechanism with metals and halogens, although it may be pericyclic in some cases, is probably more often than not ionic, with ionization and recapture as the pathway by which the halogen or metal shifts from one end of a conjugated system to the other. Secondly and more interestingly, especially with carbon, there is the possibility of sigmatropic shift with inversion of configuration in the migrating group.

To see what this means, let us begin with the [1,3] shift of a carbon atom. As with the [1,3] shift of a hydrogen atom **5.20**, the antarafacial shift of carbon **5.21** is symmetry-allowed, with the double bond taking up the role of antarafacial component; but it is just as unreasonable. The new possibility is the suprafacial shift **5.22** with inversion of configuration on carbon, with the σ-bond taking up the role of antarafacial component.

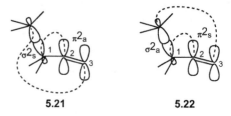

5.21   **5.22**

This was impossible with a hydrogen atom, which has only an s-orbital. This possibility, however, remains fairly remote, since the developing overlap shown in **5.22** by the long dashed line is not very plausible. Nevertheless, it appears to be the pathway taken in the [1,3] suprafacial shift **5.23** of a silyl group at 500°, with the long Si–C bonds making it that bit more reasonable. It might also be the pathway, or at least an influence on the pathway, followed in a rather constrained thermal [1,3] shift of a carbon group **5.24**→**5.25**. The deuterium atom, placed as a stereochemical marker, changes from being *trans* to the acetoxy group in the bicyclo[3.2.0]heptene **5.24** to being *cis* to it in the bicyclo[2.2.1]heptene **5.25**, showing that a remarkable inversion of configuration had taken place, in conformity to the rules. A full discussion of whether this reaction really is pericyclic is beyond the scope of this book, but we can say that, given the difficulties of achieving effective overlap in [1,3] shifts using either the σ-bond or the π-bond as an antarafacial component, [1,3] migration of carbon groups with inversion of configuration is likely to be very rare.

5.23

5.24   **5.25**

Photochemically, of course, a suprafacial shift with retention of configuration in the migrating group is allowed, and there are many examples of this reaction, such as the reversible 1,3 shift in verbenone **5.26**, but not all such reactions can be assumed to be pericyclic.

However, a suprafacial shift with inversion of configuration is not unknown even for carbon—all that is needed is a longer conjugated system to make it easier for a reasonable transition structure to be reached. A suprafacial [1,5] shift will be allowed with retention of configuration, but a suprafacial [1,7] shift will only be allowed if there is inversion of configuration in the migrating carbon **5.27**. For example, the front C–C bond of the cyclopropane in the bicyclo[6.1.0] nonatriene **5.28** migrates to C-7 to give the C–C bond (starred) at the rear in the isomeric bicyclo[6.1.0]nonatriene **5.29**. The two substituents on the migrating

5.26

5.27

carbon remain with the cyano group *exo* in the bicyclic system and the methyl group *endo*. This can be easily misunderstood as corresponding to retention of configuration, but the new bond in the product (starred) is on the back side of the migrating carbon atom, whereas the bond that broke was on the front side. This is therefore inversion as far as the migrating atom is concerned. Counter-intuitively, perhaps, it is retention of configuration that would cause the two sub-stituents to change places.

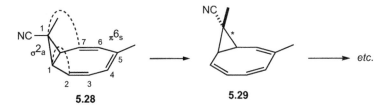

We may now summarize the rules for symmetry-allowed [1,n] sigmatropic rearrangements of carbon groups in Fig. 5.2, where *s* and *a* refer to suprafacial and antarafacial, and *r* and *i* to retention and inversion of configuration at the migrating centre. If the migrating group is hydrogen, the pathways involving inversion must be discounted.

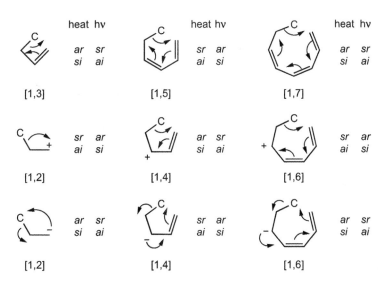

Fig. 5.2 Symmetry-allowed [1,n] sigmatropic rearrangements of carbon

It is not necessary to learn a table like this—it can be summarized with the simplified rule for thermal reactions that, if the total number of electrons is (4n + 2), a suprafacial shift with retention of configuration in the migrating centre is symmetry-allowed—as is, formally (but impossible in practice), an antarafacial shift with inversion of configuration. Reactions with (4n) electrons are the oppo-site, and photochemical reactions follow the reverse of the thermal rule. It is wise always to check any case you may come across by using the concise Wood-ward–Hoffmann rules in Chapter 3 on pp. 44 and 49, which cover everything, including longer conjugated systems than those listed here.

Fig. 5.2 includes ionic conjugated systems. The most simple and the best known is the [1,2] shift in cations **5.30**. When the migration terminus, the migration origin, and the migrating group are all carbon atoms, it is known as a Wagner–Meerwein rearrangement; when the migration terminus is a nitrogen atom it is a Curtius, Beckmann, or Lossen rearrangement; and when it is oxygen its most characteristic manifestation is the Baeyer–Villiger reaction. All of these reactions are well known to take place with retention of configuration in the migrating group, and we can now see that a [1,2] sigmatropic rearrangement with retention in the migrating group **5.31** is a symmetry-allowed $[_\sigma2_s + _\omega0_s]$ process.

In contrast, [1,2] shifts in anions **5.32** are rare, and we can now see that that is because the straightforward-looking $[_\sigma2_s + _\omega2_s]$ process **5.33** is symmetry-forbidden. They would have to follow unattainable $[_\sigma2_s + _\omega2_a]$ or $[_\sigma2_a + _\omega2_s]$ geometries to be symmetry-allowed.

Allowed but geometrically unreasonable pathways for [1,2] shifts in anions

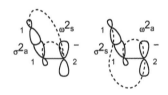

However, [1,2] shifts in anions, although rare, are known, and for some time they presented a puzzle, since the Woodward–Hoffmann rules are rarely broken. A stepwise ionic reaction did not seem reasonable, because the separated ions were not well enough stabilized. It is now clear that most such reactions, typified by the 1,2-Stevens rearrangement **5.34** → **5.36**, do take place stepwise, but by *homolytic* cleavage **5.34** → **5.35** and recombination **5.35** → **5.36**.

A [1,4] shift of hydrogen in the cyclooctenyl cation **5.37**—does not occur—it is forbidden to be suprafacial, and it is geometrically impossible for it to be antarafacial—but an allowed [1,6] shift does occur in the cyclooctadienyl cation **5.38**, where it must be, and is symmetry-allowed to be, suprafacial. This comparison was thoughtfully designed so that the distance the hydrogen atom has to move is approximately the same in the two cases, making it more plausible that the rules make the crucial difference between a reaction that is observed and one that is not.

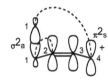

5.37

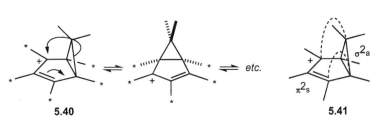

5.38

In contrast, a [1,4] suprafacial shift of carbon in an allyl cation can occur, because it is symmetry-allowed to take place with inversion of configuration **5.39**. The most striking example of this is the degenerate rearrangement of the bicyclic cation **5.40**, in which the bridge moves around the perimeter of the five-membered ring. This causes the five starred methyl groups to become equivalent on the NMR time scale, giving rise to a sharp 15-proton singlet. Meanwhile the other two methyl groups remain as two separate three-proton singlets, showing that the *endo* methyl group remains *endo* and the *exo* remains *exo*. This is the same kind of movement as we saw in the bicyclononatriene **5.28**, where inversion of configuration in the migrating group means that the substituents do not exchange places—the well-disciplined bridge atom parades round and round the ring **5.41**, obeying the rules at every turn.

5.39

5.40   *etc.*   **5.41**

Sigmatropic rearrangements in anions are more rare, but the isomerization of the conjugated anion **5.42** to the conjugated anion **5.45** probably takes place by a [1,6] shift **5.43 → 5.44**. It is known to be intramolecular, but the stereochemistry (antarafacial is allowed) could not be checked. In order for the reaction to take place at all, it is necessary for the stereochemistry within the conjugated system to change to the all-Z **5.43**, which can easily take place by the lithium counterion's reversibly forming a σ-bond to remove the contiguous overlap present in the anion itself.

5.42   **5.43**   40°   **5.44**   **5.45**

52:48

Although the stereochemistry could not be seen in the last example, it is evident in the migration of the hydrogen atom in the zwitterion intermediate **5.46**, which is symmetry-allowed as a suprafacial [1,4] shift. Although the first step is a photochemical disrotatory six-electron process, the second step is probably not photochemical, since the intermediate **5.46** would be unlikely to survive long enough to absorb another photon. Photochemical reactions often create high-energy intermediates, which then find thermal pathways for subsequent reactions.

hv

5.46   **5.47**

5.3 [m,n] Rearrangements

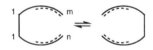

All the migrations of σ-bonds described above take place at only one end of the bond, so that all of them are [1,n] shifts. It is also possible to have the bond move m atoms along a conjugated system at the other end, as we saw in Chapter 1 with the [3,3] Claisen rearrangement **1.17→1.19**. This type of rearrangement is the most important of the [m,n] category, and is known with various atoms in the chain. The all-carbon version is known as a Cope rearrangement **5.48→5.49**, where the relief of strain in the cyclobutane ring drives the reaction forward. The presence of one oxygen makes it a Claisen rearrangement, whether it takes place in an aromatic **1.17→1.19** or an open-chain system **5.50→5.51**. The name Claisen-Cope is sometimes used to cover all [3,3] sigmatropic rearrangements, even those with sulfur or nitrogen atoms in the chain, although an important exception, with two nitrogens, is subsumed in the Fischer indole synthesis.

120°, 10 min   170°, 15 min

91%   80%

5.48   **5.49**   **5.50**   **5.51**

The reaction is symmetry-allowed when it is suprafacial on all three components, but there are two reasonably accessible transition structures for an all-suprafacial reaction, chair-like **5.52** and boat-like **5.53**, both of which are [$_\pi2_s$ + $_\sigma2_s$ + $_\pi2_s$]. In the reaction **5.48**→**5.49**, it must be boat-like to give two *cis* double bonds in the product, but this is probably constrained by the high energy of fitting a *trans* double bond into the ring. To find which was preferred in open-chain systems, the diastereoisomeric hexa-1,5-dienes *S**,*S**-**5.54** and *R*,*S*-**5.54** were separately heated. The result was that *S**,*S**-**5.54** gave mainly *E*,*E*-**5.55**, and *R*,*S*-**5.54** gave mainly *E*,*Z*-**5.55**, showing that a chair-like transition structure was preferred. Had a boat-like structure been involved, *S**,*S**-**5.54** would have given *E*,*Z*-**5.55**, and *R*,*S*-**5.54** would have given *E*,*E*-**5.55**.

5.52   5.53

In the absence of constraints, chair-like transition structures are preferred to boat-like transition structures.

S*S*-**5.54**   EE-**5.55**   RS-**5.54**   EZ-**5.55**

This conclusion has been applied in the Ireland–Claisen rearrangement **5.56**→**5.58**, which is one of the most frequently used [3,3] sigmatropic rearrangements, because it sets up the relative configuration of two usefully substituted stereogenic centres with high levels of predictability, stemming from the chair-like transition structure **5.57**.

5.56   5.57   5.58

The oxyanion of the enolate (or a silyloxy group) appears to speed up this reaction relative to an unadorned Claisen system, but an even more dramatic acceleration is seen when Cope systems have an oxyanion substituent on the tetrahedral carbon, as in the reaction **5.59**→**5.61**, which takes place at a much lower temperature than normal Cope rearrangements, and is usually called an oxy-Cope rearrangement. This rearrangement also serves to show how different the starting material **5.59** and the product **5.61** can look.

5.59   5.60   5.61

A helpful trick: draw the product **5.60** without changing the position of any of the atoms; just change the bonds dictated by the curly arrows. It is then easier to see how a structure drawn like **5.59** can give a product drawn like **5.61**.

In the Claisen rearrangement **5.62**→**5.63** the product cannot aromatize by loss of a proton, as in the earlier example **1.17**→**1.19**. Instead a Cope rearrangement takes place to give a *para*-substituted phenol **5.64**. A longer conjugated system **5.65** allows a more direct delivery to the *para* position, giving the phenol **5.67** as the major product along with some of the product of a normal Claisen rearrangement. The terminal methyl group in **5.62** acts as a label, showing that a somersault of two [3,3] sigmatropic rearrangements is taking place, and not a direct [3,5] reaction. Similarly, the terminal methyl group in **5.65** shows that this is a [5,5] rearrangement **5.65**→**5.66** rather than two successive [3,3] rearrangements. The [5,5] rearrangement is allowed if it is all-suprafacial in a geometry **5.68** that is not difficult to achieve. The *para*-benzidine rearrangement is also a [5,5] sigmatropic rearrangement.

	[3,3]		[3,3]		
	Claisen		Cope		
	186°				
	3.5 h			91%	

5.62   **5.63**   **5.64**

	[5,5]		
	185°		
	3.5 h	50%	

5.65   **5.66**   **5.67**   **5.68**

The world record at the moment is a [9,9] bisphenylogue **5.69**→**5.70** of the benzidine rearrangement.

5.69   **5.70**

[2,3]-Sigmatropic rearrangements **5.71**, like the Mislow racemization of sulfoxides **1.22** in Chapter 1 on p. 6, have many variants, depending upon which atoms are present in the chain of five atoms. There is an all-carbon example, but most have Y = O, and there are well-known families of reactions with X = O, N, and S. With X = O and Y = C, it is called a [2,3]-Wittig rearrangement, as in the reaction of the bisallylic ether **5.72** with strong base. The base selectively deprotonates the allylic ether **5.72** at the less substituted of the allyl carbons next to the oxygen, and the anion **5.73** rearranges to give largely the alcohol **5.74** with the two substituents *anti* on the carbon chain.

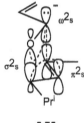

5.71

5.72   BuLi   **5.73**   [2,3]
75%   **5.74**

This rearrangement obeys the Woodward–Hoffmann rule, illustrated as an allowed $[_\sigma 2_s + _\omega 2_s + _\pi 2_s]$ reaction **5.75**. The likely transition structure has an envelope conformation, allowing the σ-overlap to develop properly head on and the π-overlap sideways on. One of the dashed lines in **5.75**, the one between the oxygen atom and the carbanion centre, is a bit different from those we have seen before—it identifies the carbanion centre as conjugated to the σ-bond in the starting material. This means that all the orbitals are conjugated in the cyclic transition structure, making the reaction pericyclic, even though the curly arrows in **5.71** and **5.73** do not appear to complete a circle.

Similarly, with X = N or S. The quaternary ammonium salt **5.76** is deprotonated adjacent to the ester group to give the ammonium ylid **5.77**, which rearranges with ring expansion to give a nine-membered ring with a *trans* double bond **5.78**. Intramolecular alkylation of the sulfide **5.79** followed by deprotonation gives a sulfonium ylid **5.80**, which rearranges with ring expansion to give a cyclodecanone with a sulfide bridge **5.81**.

5.75

5.76   DBU   **5.77**   [2,3]
90%   **5.78**

5.79   K$_2$CO$_3$ **5.80**   [2,3]
93%   **5.81**

The transition structure **5.75** also shows some of the constraints that make these reactions useful methods for controlling stereochemistry. Although not true of every [2.3]-Wittig rearrangement, the preferred envelope conformation places the isopropyl group in a pseudo-equatorial position, and the anion-stabilizing vinyl group *exo*, away from the fold in the envelope. The result **5.74** is a *trans* double bond with the methyl and hydroxyl groups *anti* on the carbon backbone, but if the double bond had been *cis*, it would have led to the methyl and hydroxyl groups being *syn*, with both groups behind the chain. The reaction of the ammonium ylid **5.77** is already a slightly different case, because the anion-stabilizing group is *endo* in the transition structure, but it then appears in a pseudo-equatorial position in the product **5.78**.

Mislow's [2,3]-sigmatropic rearrangement of sulfoxides is more than a mechanistic curiosity, because the intermediate sulfenate **5.83** can be intercepted by a suitably thiophilic reagent, converting an enantiomerically enriched sulfoxide **5.82** into a comparably enriched rearranged allyl alcohol **5.84**, with suprafacial shift **5.85** of the functionality.

| 5.82 | 5.83 | 5.84 | 5.85 |

There are many variants on these reactions, including the [3,4] shift in the cation **5.87** created by protonating the ketone **5.86**. The allylic shift is proved by labelling the carbon marked with an asterisk using deuterium and showing that the deuterium in the product **5.88** is at the benzylic position.

| 5.86 | 5.87 | 5.88 |

There are also vinylogues of the [2,3] rearrangement like the [4,5] shift in the ammonium ylid **5.90**. This intermediate had available a [2,3] shift just like that in the ylid **5.75**; but the sigmatropic shift takes place predominantly through the longer system. The other possibilities, a [2,5] or a [3,4] shift, are forbidden to be suprafacial on both components. A transition structure with one antarafacial component might be accessible, given the length of the conjugated systems, but it does not compete with the all-suprafacial pathway.

5.89   **5.90**   **5.91**

It is frequent but not invariable that where a longer conjugated system has a geometrically accessible and symmetry-allowed transition structure like that in **5.90**, the longer system is used rather than one of the shorter systems. Thus, the [8 + 2] and [6 + 4] cycloadditions in Chapter 2 on p. 17, and the [14 + 2] cycloaddition in Chapter 3 on p. 49 take place rather than perfectly reasonable Diels–Alder reactions, and the eight-electron electrocyclic reactions of **4.51** and **4.54** in Chapter 4 on p. 71 take place rather than the disrotatory hexatriene-to-cyclohexadiene reactions that are available. This kind of selectivity is called **periselectivity**.

5.4 Further reading

Claisen and Cope rearrangements:

S. J. Rhoads and N. R. Raulins, *Org. React. (NY)*, 1975, **22**, 1; several chapters on this and related reactions in *COS*, Vol. 5, ed. L. A. Paquette: R. K. Hill, Ch. 7.1; P. Wipf, Ch. 7.2; F. E. Ziegler, Ch. 7.3; E. Piers, Ch. 8.2.

[2,3]-Sigmatropic rearrangements:

R. Brückner, Ch. 4.6 in *COS*, Vol. 6, ed. E. Winterfeldt.

A special case:

a [5.5]-sigmatropic rearrangement visualized: http://www.ch.imperial.ac.uk/rzepa/blog/?p=10252

5.5 Problems

5.1   The following reactions take place with one or more sigmatropic rearrangements. Identify the reactions, and show that they obey the Woodward–Hoffmann rule.

(a)

(b)

(c)

5.2 Identify the steps and predict the stereochemistry at the wavy lines of each of these two-step reactions:

(a)

heat

(b)

heat

(c)

heat

5.3 The obvious Diels–Alder cycloadditions in these two cases did not take place. Identify the pathways that led them astray, and explain why the reactions observed competed with the direct reaction.

(a)

—OSiMe$_3$ +

CO_2Me

CO_2Me

100°, 30 h

Me$_3$SiO

CO_2Me

CO_2Me

(b)

+

Cl⁀CN

r.t.

Cl

CN

5.4 Identify the sigmatropic rearrangements that take place, in addition to other steps, in these reactions:

(a)

1. [pyrrolidine]

2. H$_2$O

(b)

650 °C

5.6 Summary

- Sigmatropic reactions are unimolecular rearrangements of two common kinds: a [1,n]-rearrangement, in which a hydrogen atom, or more rarely a carbon atom, migrates n atoms along a conjugated chain; or an [m,n]-rearrangement when one end of a σ-bond migrates m atoms along one chain and the other end migrates n atoms along the other chain.

- When the total number of electrons involved is a number that can be described in the form (4n + 2), the symmetry-allowed reaction can be drawn as an all-suprafacial event.

- When the total number of electrons involved is a number that can be described in the form (4n) then the symmetry-allowed pathway must be drawn with one component antarafacial and the other (or others) suprafacial.

- The migration of a single bond can take place by leaving one surface of a conjugated system and arriving on the same surface at the other end, m or n atoms away. Such migrations are described as being *structurally* suprafacial. Thus [1,5]-shifts of hydrogen take place suprafacially in the structural sense.

- Alternatively, the migration of a single bond can take place by leaving one surface of a conjugated system and arriving on the opposite surface m or n atoms away. Such reactions are described as being *structurally* antarafacial. Thus, the [1,7]-shift of hydrogen takes place antarafacially in the structural sense.

- When the migrating group in a [1,n]-rearrangement is carbon, and when the total number of electrons involved is a $(4n + 2)$ number, the structurally suprafacial shift will take place with retention of configuration in the migrating group. The easy [1,2]-shift in cations (as in Wagner–Meerwein, Beckmann, Curtis, and Baeyer–Villiger reactions) are examples of this pathway.

- When the migrating group in a [1,n]-rearrangement is carbon, and when the total number of electrons involved is a $(4n)$ number, a structurally suprafacial shift can take place with inversion of configuration in the migrating group when an accessible transition structure is available.

- [1,2]-Shifts in carbanions have no readily accessible pericyclic transition structure, and are usually stepwise reactions by way of radical intermediates.

- The most common [m,n]-rearrangements are [3,3]-Claisen–Cope rearrangements and [2,3]-rearrangements like the Mislow rearrangement of allylsulfoxides and the 2,3-Wittig rearrangement of the anions of allyl ethers.

- The Claisen–Cope rearrangement is allowed to use either a boat-like or a chair-like transition structure, with the chair-like usually the lower in energy.

- Longer [m,n] systems are known, and when available are usually faster than the shorter, a form of selectivity known as periselectivity.

Group transfer reactions

6.1 Introduction

6.1   **6.2**

In Diels–Alder reactions **6.1** only π-bonds are broken, allowing the existing and the new σ-bonds to complete a ring in the product **6.2**. If, on the other hand, one of the π-bonds in the diene is replaced by a σ-bond to a hydrogen atom, then a similar-looking cyclic reaction can take place **6.3** (arrows), but no ring is formed in the product **6.4**. Similarly, if both π-bonds in the diene are replaced by σ-bonds to hydrogen **6.5**, or the π-bond of the dienophile is replaced by a σ-bond between two hydrogen atoms **6.7**, no ring is formed in the corresponding products **6.6** and **6.8**. In all three cases one of the components suffers an addition reaction, and at least one hydrogen atom is transferred from one molecule to the other. These are the common six-electron **group transfer reactions**.

6.3   **6.4**   **6.5**   **6.6**   **6.7**   **6.8**

The reaction **6.3** → **6.4** is called an ene reaction—named thus because it is like the diene reaction but with only one double bond. When the atom transferred is a metal atom, the reaction is called a metalla-ene reaction. It is common, but is not usually discussed as a pericyclic reaction.

Just as there are hetero-Diels–Alder reactions, there are hetero versions of group transfer reactions. In Chapter 1, on p. 1 the purely thermal aldol reaction of an enol and a ketone is identified as a hetero group transfer reaction. There are group transfer reactions using larger numbers of electrons, but they are not common, or not commonly recognized as pericyclic reactions.

6.2 Ene reactions

With one σ-bond to break, ene reactions are not in general as easy as Diels–Alder reactions. Typically an ene reaction will require temperatures 100° or more higher than a comparable Diels–Alder reaction. As with Diels–Alder reactions, electron-withdrawing groups on the enophile speed up the reaction and are almost obligatory in bimolecular reactions. Equally electron-donating groups on the ene helpfully speed up the reaction. The regiochemistry with an unsymmetrical enophile **6.9** is such that the major product **6.10** has the alkene carbon attacking the β-position of the enone system and the hydrogen atom going to the α-position. Lewis acid catalysis **6.11**→**6.12** makes the reaction even more amenable.

6.9　　　　　　　　　　　**6.10**　　R = H　7:1; R = C$_5$H$_{11}$-n　3:1

6.11　　　　　　　　　　　**6.12**

Some aspects of the stereochemistry of attack are revealed by the reaction **6.14** + **6.15**→**6.16**. The symmetry-allowed, six-electron, all-suprafacial transition structure **6.13** has an envelope conformation, with the two new σ-bonds developing overlap head on, and with the enophile oriented, just as in a Diels–Alder reaction, so that the electron-withdrawing group is *endo*−lying over the fold of the envelope. The result is that if the enophile has an α-substituent and the alkene a substituent on C-3, it is possible to control the relative configuration of a pair of 1,3-related stereocentres (**6.16** asterisks). Notice that this type of stereocontrol is not governed by the Woodward–Hoffmann rules themselves, but by secondary effects imposed on the all-suprafacial reaction. The presence of the alkyl group R on C-2 causes the alkene to react at C-3, making the geminal methyl group C-1 deliver the hydrogen atom rather than the vicinal methyl group C-4.

6.13

6.14　　　　　　**6.15**　　　　　　　　**6.16**

The ene reaction has proved to be particularly powerful in synthesis when carried out intramolecularly. The usual increase in rate for an intramolecular reaction allows even unreactive partners like hydrocarbons to combine. Thus, the diene **6.17** gives largely (14:1) the *cis* disubstituted cyclopentane **6.19** by way of a transition structure **6.18**. It is important to recognize that the selective formation of the *cis*-disubstituted cyclopentane has nothing to do with the rules for pericyclic reactions. It is a consequence of the lower energy when the trimethylene chain spans the two double bonds in such a way as to leave the hydrogen atoms on the same side of the folded bicyclic structure. This constraint is related to the way a five-membered ring fused to a five- or six-membered ring is lower in energy if it is *cis*-fused.

6.17   6.18   6.19

The chain of atoms can include an oxygen atom. When the enophile is a carbonyl group, the reaction **6.20** is called a Prins reaction. Prins reactions are usually acid-catalysed with the proton transfer step preceding the C–C bond-formation. They are not then pericyclic, but a gas phase reaction like the intramolecular example **6.21**→**6.23**, which has a plausible transition structure **6.22**, is probably pericyclic. The acid-catalysed reaction of this substrate gives a completely different product.

6.20

6.21   6.22   6.23

There are other ene-like reactions with an oxygen atom in the chain. Thus, an enol can deliver a hydrogen atom from oxygen **6.24** in a purely thermal reaction with an alkene, although the temperature has to be high. This reaction is called a Conia reaction, and is most useful when intramolecular. Since the enol in an unsymmetrical system cannot usually be made specifically, this reaction cannot be regiocontrolled–it depends upon one of the two or more possible enols' reacting in the desired way. When that does happen, it is an effective synthetic method, giving, for example, camphor **6.27** from dihydrocarvone **6.25** in one operation, by way of the enol **6.26**.

6.24

6.25   6.26   6.27

6.3 Retro ene reactions and other thermal eliminations

All-carbon ene reactions can go in reverse when ring-strain is released, as in the vinylcyclopropane **6.28**. This reaction is curious, because it bears some similarity to a [1,5]-sigmatropic rearrangement. It is a homologue of that reaction, since it is quite common to think of the chemistry of cyclopropanes as similar to, but usually less favourable than, that of alkenes.

6.28

With heteroatoms in the chain it is possible to drive such reactions in reverse without having to release strain. Thus, esters such as acetates and benzoates **6.29** undergo a cyclic β-elimination on pyrolysis, in what amounts to a retro-ene reaction. The temperatures needed are rather high, typically about 400°, and the reaction is best carried out by flash vacuum pyrolysis with short contact times to minimize bimolecular side reactions. This type of elimination is known to be *syn* stereospecific, as it must be to follow the all-suprafacial pathway.

6.29

The Chugaev reaction **6.30**→**6.31** + **6.32**, using a xanthate ester in place of the acetate or benzoate, is relatively fast and convenient, requiring temperatures typically of 150–250°.

6.30 **6.31** 45% **6.32** 32%

An even milder cycloelimination uses a ring of five atoms **6.33** instead of six, but still involves six electrons. This is no longer a retro-ene reaction, but it is still a retro group transfer; and it is allowed in the all-suprafacial mode **6.34**. The pyrolysis of *N*-oxides **6.35** is called the Cope elimination, and typically takes place at 120°. The corresponding elimination of sulfoxides **6.36** (X = S) typically takes place at 80–100°, and, even easier, the elimination of selenoxides (X = Se) takes place at room temperature or below. All these reactions are reliably *syn* stereospecific.

6.33

6.34

6.35 **6.36**

The rates of these reactions are affected by functionality making the temperatures quoted only rough guides. Thus, better electron-withdrawing groups than the phenyl group in **6.36** attached to the hydrogen-bearing carbon speed up the reaction **6.37**→**6.39**, showing that the transition structure **6.38** must have substantial negative charge built up on that carbon.

6.37	**6.38**	**6.39**

6.4 Diimide and related reductions

The transfer of two atoms of hydrogen from carbon atoms **6.5** in a *syn* stereospecific manner is clearly allowed as a $[_\sigma2_s + 2_s + 2_s]$ reaction **6.40**. The difficulty is that two σ-bonds must break at the same time, a heavy penalty very few organic reactions can overcome. However, it is known in a handful of cases, requiring, for example, the gain, twice over, of aromaticity in the reaction of 9,10-dihydronaphthalene **6.41** with dimethylcyclohexene **6.42** to give naphthalene **6.43** and *cis*-dimethylcyclohexane **6.44**. The reaction is also known in a few cases helped by the high degree of organization found in intramolecular reactions.

6.41	**6.42**	**6.43**	**6.44**

The 10-electron bisvinylogue, in which 1,4-cyclohexadiene **6.45** reduces anthracene **6.46** to give benzene **6.47** and dihydroanthracene **6.48**, is also known. Although this ten-electron event is clearly allowed in the all-suprafacial mode, deuterium labels show that the delivery of the two hydrogen atoms is not stereospecifically *syn*; it is not therefore pericyclic. This is perhaps a timely reminder that not all reactions that appear to obey the rules can be assumed to be pericyclic.

6.45	**6.46**	**6.47**	**6.48**

Diimide reduction **1.28** in Chapter 1 on p. 6, with diimide generated by the oxidation of hydrazine and by other methods like that in Chapter 2 on p. 11, is much the most common reagent used to transfer two hydrogen atoms in a pericyclic reaction. The reaction is made more favourable by the formation of the low-energy molecule dinitrogen. Diimide reduction works well with double and triple bonds that are not too hindered; diimide is not poisoned by sulfur-containing compounds as hydrogenation catalysts are; and it is highly selective for homonuclear double bonds, C=C and N=N, over heteronuclear double bonds like C=O. The selectivity probably comes about because the pericyclic nature of the reaction sets a premium on symmetry. Its *syn* stereospecificity certainly does, since it corresponds to the allowed all-suprafacial pathway **6.40**. Its pericyclic nature, however, makes it unsuitable for reduction with deuterium, because the isotope effect for both C–D bonds breaking in concert is too high.

6.5 1,4-Elimination of hydrogen

The addition of dihydrogen across the termini of a diene, drawn on p. 98 as **6.7**, is allowed as a $[_\pi 4_s + _\sigma 2_s]$ reaction **6.49**. It is only seen in reverse as a 1,4-elimination of hydrogen **6.8** → **6.7**, as in the relatively easy thermal aromatization of 1,4-dihydrobenzene to benzene **6.50** → **6.51**. The reaction is known to be stereospecific, with the two atoms *cis* to each other on the cyclohexadiene forming the dihydrogen molecule.

6.49

6.8   **6.7**   **6.50**   **6.51**

In contrast, the aromatization of 1,2-dihydrobenzene requires a much higher temperature, typically 400°. It is unlikely to be a pericyclic reaction—the only geometrically reasonable transition structure is for *syn* elimination with a forbidden all-suprafacial four-electron process; the allowed anti elimination, with one component antarafacial and the other suprafacial, is geometrically unreasonable. It is most likely a stepwise and radical-induced reaction.

6.6 Further reading

Ene reactions:

H. M. R. Hoffmann, *Angew. Chem. Int. Ed. Engl.*, 1969, 8, 556.

W. Oppolzer and V. Snieckus, *Angew. Chem. Int. Ed. Engl.*, 1978, 17, 556/

B. B. Snider in *COS*, Vol. 5, Ch. 1.1.

Eliminations: P. C. Astles, S. V. Mortlock, and E. J. Thomas, in *COS*, Vol. 6, Ch. 5.3.

6.7 **Problems**

6.1 Identify the two group transfer reactions used in the synthesis **6.52** → **6.53** of a chiral methyl group, and show that they obey the Woodward–Hoffmann rule.

6.2 Identify the two pericyclic steps in this reaction.

6.8 **Summary**

- Group transfer reactions are similar to cycloadditions, except that one of the π-bonds is replaced by a σ-bond, usually to hydrogen. In consequence an atom is transferred from one molecule to the other and no ring is formed.

- The most common group transfer reaction is the ene reaction between one alkene having an allylic hydrogen and another alkene equipped with an electron-withdrawing group, transferring the hydrogen atom to one end and an allyl group to the other.

- Electronegative heteroatoms, usually O, may replace any of the carbon atoms in an ene reaction.

- The thermal β-elimination of esters, N-oxides, sulfoxides, and selenoxides are reverse group transfer reactions. Involving six electrons, they are *syn* eliminations taking place with all-suprafacial transition structures.

- A second common group transfer reaction is the *syn* delivery of two hydrogen atoms from the reactive intermediate diimide to a C=C or N=N double bond.

- A third, but more rare, group transfer reaction is the loss of dihydrogen from 1,4-dihydroaromatic rings.

Exercises

The problems given at the end of each of the preceding chapters were restricted to exercises with only the kind of pericyclic reaction covered by that chapter and any preceding it. These exercises are drawn from the whole range of pericyclic reactions, together with ionic reactions of many kinds, which may precede or follow them.

E.1 The following reactions use two or more pericyclic steps. Identify all the reactions, and draw reasonable looking transition structures obeying the Woodward–Hoffmann rule.

E.2 Explain the formation of the abnormal Claisen product **E.3**, which is known to be produced from the normal product **E.2**.

E.1   E.2   E.3

E.3 The tetracyclic natural product **E.5** can be prepared from the linear polyene **E.4** at 100°. This is also believed to be the path followed in its biosynthesis. All the steps are pericyclic. What are they?

E.4 Identify the two and the three pericyclic steps involved in the formation of the major and minor products **E.6** and **E.7**, respectively, in this transformation (hint: all four classes of pericyclic reaction are represented):

E.5 Identify the pericyclic steps in these reactions:

(a)

(b)

(c)

E.6 Identify how and why the aldehydes in these two reactions give the products **E.8** and **E.9**, respectively, the first catalytic in aldehyde and the second stoichiometric, and explain the stereochemistry of the products, both of which retain the enantiomeric purity of the starting materials:

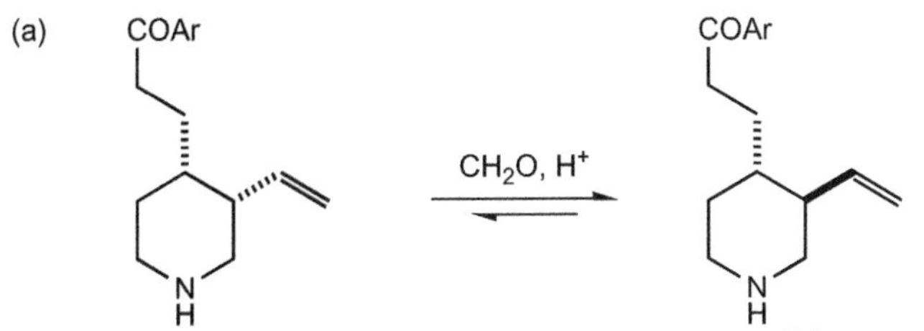

E.7 This complicated looking reaction is simply two Diels–Alder reactions. Identify them, and suggest which took place first and why?

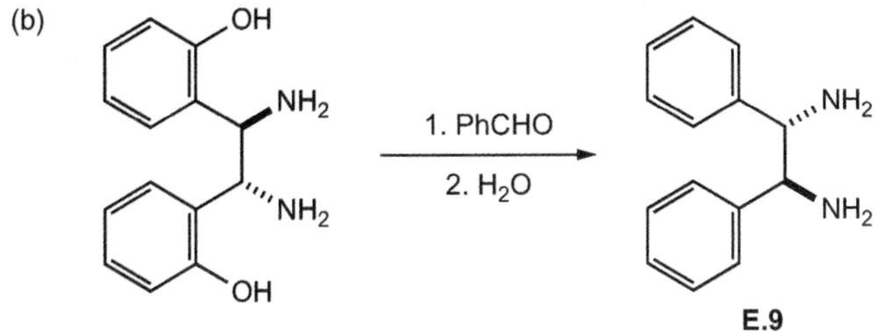

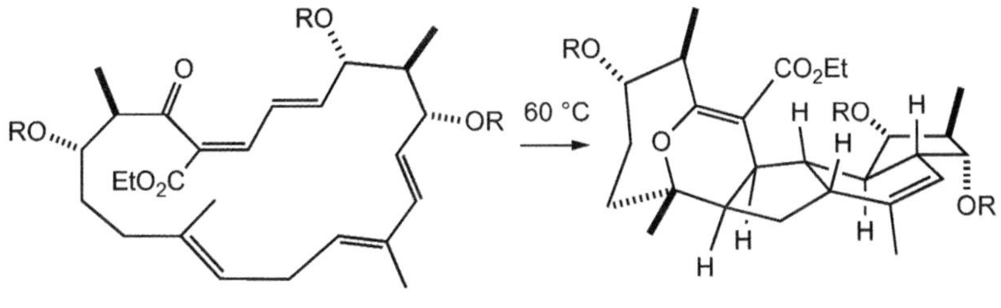

E.8 This sequence of reactions uses five pericyclic steps, as well as a stereospecific *syn* hydrogenation. What would be the geometry of the deuterium-labelled cyclohexadiene?

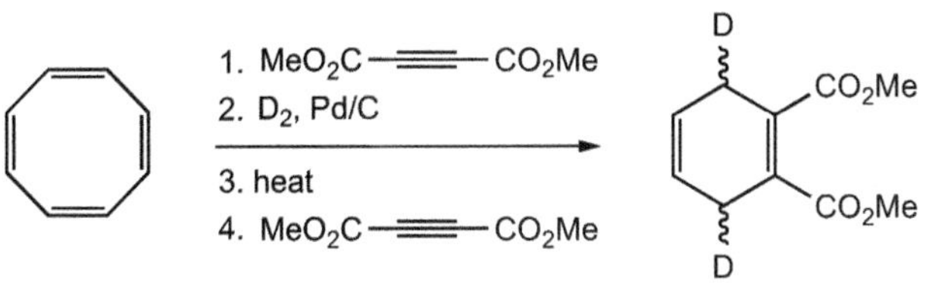

To keep the answer section short and concise, the following codes are used:

D–A = Diels–Alder; EC = electrocyclic closing; EO = electrocyclic opening; GT = group transfer; Ch = cheletropic; Cy = cycloaddition; 1,3-D = 1,3 dipolar cycloaddition; con = conrotatory; dis = disrotatory; s = suprafacial.

Answers to Problems

Chapter 1

1.1a D–A→cyclohexa-1,4-diene; –H_2 by retro-GT (also arguably classifiable as a retro-Cy).

1.1b EO of a cyclobutene, EC of a hexatriene.

1.1c GT of $2H_2$ to give 2 *cis* double bonds, 1,5-H shift from the preferred conformation **A.1** makes one *trans*.

1.1d Retro-Ch loss of CO; retro-D–A.

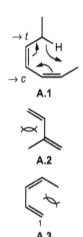

A.1

A.2

A.3

Chapter 2

2.1 2-Me raises energy of s-*trans* conformation **A.2** bringing it closer in energy to that of the s-*cis* diene, raising the proportion of s-*cis* at equilibrium, and hence increasing the rate of cycloaddition. The 4-Me group *cis* to the double bond in piperylene **A.3** reduces the proportion of diene in the s-*cis* conformation, because of steric compression with C-1, and hence decreases the rate of D–A.

2.2a 1,3-D.

2.2b Allyl cation + diene.

2.2c After an H shift from O-to-O, pentadienyl cation + alkene intramolecularly.

2.3 D–A of C≡N across the 3,6 positions; retro-D–A with loss of N_2; D–A of C≡C across the 2,5 positions; retro-D–A with loss of PhC≡N.

2.4 D–A of p-benzoquinone to cyclohexa-1,3-diene and intramolecular [2 + 2] Cy hv; 2 × D–A of butadiene to acetylenedicarboxylate and intramolecular hv.

2.5 Antiaromaticity of cyclobutadiene is lost twice over in one step, and the orbitals are spaced better for good overlap in a D–A to develop (C-1 and C-4 of an open-chain diene are further apart).

Chapter 3

3.2 A two-step ionic reaction competes with a D–A; the former is faster in the more polar solvent. The two-step reaction takes place on the more abundant s-*trans* conformer, and the intermediate allyl cation is fixed in an extended configuration so that only the cyclobutane can be formed from it, not the cyclohexene.

3.3 [4 + 4] hv like **2.81**, retro-D–A. D–A, retro-D–A. 1,3-D, retro-D–A of CO_2, intramolecular 1,3-D.

3.4 Base removes benzylic H, retro-'allyl'-anion + alkene (where the 'allyl' anion is the carboxylate ion) gives *trans*-cyclooctene. This is an exercise

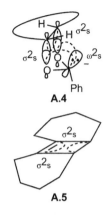

A.4

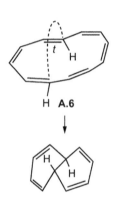

A.6

in drawing believable transition structures like **A.4**. Retro-cycloaddition, $[2_s + 2_s]$ on both σ-bonds **A.5**, makes a *trans* double bond.

3.5 The oxyanion pushes out the Cl⁻, with help from the Lewis acid, to give a cationic oxa-aza-diene (a vinylnitrosyl cation), which undergoes a hetero D–A to the alkene; the base removes the proton α to the imino cation in the product, and a retro-hetero-D–A gives the final product. The oxime N adds to the β position of the acrylate, the proton on the oxygen is removed (possibly intramolecularly by the enolate) to give a nitrone, which undergoes 1,3-D to another molecule of acrylate.

Chapter 4

4.2 EC con and EO con; as long as both steps are strictly con, the isomer **4.122** cannot be formed.

4.3 A ketone is cation-like, so treat **4.123** as a cyclopropyl cation (with a highly stabilizing oxyanion substituent): EO dis and EC dis in the opposite sense gives the enantiomer; as long as both steps are strictly dis, the isomer **4.125** will not be formed.

4.4a EO con (with the Me group outward and the vinyl group inward) and EC dis → di-Me *cis*.

4.4b EO con gives a cyclodecapentaene **A.6** with one *trans* double bond; EC dis (using the *trans* double bond, dashed line) → *trans* ring junction.

4.4c EO con → *E,E*-diene, D–A with *endo* TS → 'para' all-*cis* adduct.

4.4d EC con → the two Hs *trans* to each other.

4.4e Epoxide is isoelectronic with cyclopropyl anion, 4-electron EO con → U-shaped 'allyl' anion-like system (alternative W-shaped anion cannot proceed to product), six electron EC dis → Ph and Me groups *trans*.

4.5 N lone pair attacks C=O from less-hindered side (opposite the MeO group) without changing configuration of C=N, EC con gives the *cis* β-lactam.

4.6a EO dis (photo) and thermal 1,3-D $[4_s + 2_s]$.

4.6b EC dis → bicyclo[4.2.0]octatriene, D–A.

4.6c Ch cycloaddition of carbene to central double bond (with highest coefficients in HOMO), EO with departure of one of the two Cl as an anion.

4.6d **4.51** → **4.53**, D–A, retro D–A, and EO con.

Chapter 5

5.1a Mislow and retro-Mislow, the sulfoxide somersaults from right to left.

5.1b LDA and TBDMSCl make silyl enol ethers, then two suprafacial Ireland–Claisen steps.

5.1c Claisen then 2 × Cope, chair TS → *trans* double bonds.

5.2a EO of cyclobutene, [1,5]-H shift → *cis* double bond.

5.2b Chair Cope → *cis*-1,2-di-*trans*-1′-propenylcyclobutane, boat Cope → Me groups *cis*.

5.2c [1,5]-Shift of C, s with retention → leaves Me group in front, then [1,5]-H-shift within the cyclopentadiene ring moves the double bonds so that neither is exocyclic to the six-ring.

5.3a [1,5]-H-shift followed by a second brings the double bonds into conjugation with the silyloxy group, making it a more reactive diene for the D–A reaction.

5.3b [1,5]-H-shift brings the double bonds into conjugation with the methyl group, making it a more reactive diene for the D–A reaction.

5.4a The ketone group forms a pyrrolidine enamine towards the methyl group, Claisen followed by hydrolysis of the enamine gives the diketone.

5.4b EO, [1,5]-H-shift and EC.

Chapter 6

6.1 Intramolecular ene delivers the H; retro-ene delivers the D, O_3 → acetate.

6.2 Ene creates a conjugated diene, intramolecular D–A.

End of book exercises

E.1 Claisen → aldehyde; two possible ene reactions: side-chain ene and ring ene. Claisen to furan 2-position, and Cope restoring the furan. [1,5]-C shift, [1,5]-H shift, D–A.

E.2 Ene like **6.26** on **E.2** makes cyclopropane, then the opposite retro-ene on the Et group gives the disubstituted alkene.

E.3 EC con of octatetraene unit; EC dis of resultant hexatriene (like **4.52** → **4.53**); intramolecular D–A, s on both diene and dienophile.

E.4 EC dis of hexatriene → norcaradiene, and D–A gives **E.6**; ene of cycloheptatriene with acetylene, then EC dis of new hexatriene → new norcaradiene, then Cope opens the cyclopropane to give **E.7**.

E.5a Acid-catalysed enol fomation, EC dis of hexatriene, D–A, retro-aldol-like cleavage, dienol formation, and elimination of malondinitrile.

E.5b 1,3-D, β-elimination gives minor product; major product begins with Claisen-like rearrangement giving **A.7**, which opens by ionization of the cyano- and ketone-stabilized carbanion and subsequent loss and gain of H^+.

E.5c EO, D–A like **2.146** → **2.148**, loss of H_2 by retro GT.

E.6a Formaldehyde reversibly forms an iminium ion with the piperidine, this undergoes [3,3]-sigmatropic rearrangement, and then reverses on the other side, and the iminium ion step reverses to give the lower-energy *trans* product **E.8**.

A.7

E.6b Benzaldehyde forms imines with both amino groups, the bis-imine undergoes [3,3]-sigmatropic rearrangement with chair TS, and the imine undergoes hydrolysis; the o-hydroxy groups conjugated to the imines makes the product imine lower in energy than the starting imine.

E.7 D–A between the diene at the bottom right and the γ,δ-double bond of the enone system is an activated pair, since the γ,δ-double bond effectively carries an electron-withdrawing group, the two ends are close forming a five-ring and the diene unit comes down over the γ,δ-double bond to be suprafacial on both components; the hetero D–A is then also forming a 5-ring with the enone system coming in from below; note that the α,β-double bond must change its configuration from Z to E before the hetero-D–A.

E.8 EC dis, D–A, selective *syn* hydrogenation of the strained cyclobutene double bond, retro-D–A gives 3,4-*cis* dideuteriocyclobut-1-ene, EO con gives the E,Z-butadiene, D–A puts the deuterium atoms *trans* to one another.

Glossary

Antarafacial (a) applied to the component of a pericyclic reaction (or drawing of a component of a pericyclic reaction) describes the situation in which the new bonds are formed (or drawn) on opposite surfaces of the conjugated system: **G.1** or **G.2** for a conjugated system of π-bonds, in which the dashed lines represent the direction from which the new bonds are forming, and **G.3** or **G.4** for a σ bond. The opposite is suprafacial (qv.).

Antarafacial (b) applied to a sigmatropic rearrangement describes the *structural* change in which a substituent R attached to a σ bond breaks off from one surface of a conjugated system and forms a σ bond on the opposite surface at the other end **G.5**.

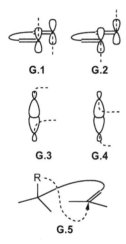

| G.1 | G.2 |
| G.3 | G.4 |

G.5

C-substituent is a substituent extending conjugation, like vinyl or phenyl, without necessarily contributing in the way of electron donation or electron-withdrawing power.

Cheletropic reactions are a special group of cycloadditions or cycloreversions in which two σ-bonds are made or broken to the *same* atom.

Components (a) in descriptions of pericyclic reactions are the set of atoms with the conjugated π- or σ-bonds (the core electronic systems) that undergo a change in the course of the reaction.

Component (b) can also mean the whole molecule taking part in a reaction, with context making it clear whether the meaning (a) or (b) applies.

Concerted reactions are those in which two or more bonds are being made or broken at the same time; a sub-class of concerted reactions are those that are synchronous (qv.).

Conrotatory electrocyclizations are those in which the rotations about the terminal double bonds are taking place either clockwise at both ends **G.6** or anticlockwise at both ends **G.7**. The term can equally be applied to electrocyclic ring openings.

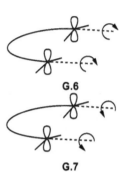

G.6

G.7

Correlation diagrams identify, using symmetry descriptors, which molecular orbitals (or states) in the starting materials become which molecular orbitals (or states) in the products.

Cycloadditions are characterized by two components' coming together to form two new σ-bonds, at the ends of the conjugated orbitals of both components, joining them together to form a ring, with a reduction in the length of the conjugated system of orbitals in each component.

Cycloreversion the reverse of a cycloaddition—a ring breaks into two components.

Dienophiles are the two-electron components of a Diels–Alder reaction, typically alkenes or alkynes having electron-withdrawing substituents.

1,3-Dipolar cycloadditions are cycloadditions between 1,3-dipoles (qv.), and alkenes, alkynes, or other π bonds like carbonyl groups and imines.

1,3-Dipoles are conjugated systems of three atoms, X, Y, and Z, each carrying a p-orbital and with a total of four π electrons in the conjugated system; X, Y, and Z are commonly almost any combination of C, N, O, and S, with a double or, in those combinations that can support it, a triple bond between two of them in the standard Lewis structure.

Dipolarophiles analogous to dienophiles (qv.), are the components of 1,3-dipolar cycloadditions (qv.) that react with 1,3-dipoles (qv.); they have a double or triple bond between any pair of atoms A and B of the common elements C, N, O, and S.

Disrotatory electrocyclizations are those in which the rotations about the terminal double bonds are taking place clockwise at one end but anticlockwise at the other **G.8**. The term can equally be applied to electrocyclic ring openings.

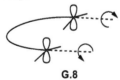

G.8

Electrocyclic reactions are characterized by the creation of a ring from an open-chain conjugated system, with a σ-bond forming across the ends of the conjugated system, and with the conjugated system becoming shorter by one p-orbital at each end.

Endo rule for Diels–Alder reactions, established by Alder, identifies as more favourable those reactions that have an *endo* transition structure, in which the electron-withdrawing substituent on the dienophile is directly under (or over) the diene's π system. It usually leads to the less-stable (because stereochemically more crowded) of the two possible stereoisomeric adducts.

Frontier orbitals are the highest occupied molecular orbital (HOMO) and the lowest unoccupied molecular orbital (LUMO).

Group transfer reactions resemble cycloadditions in which one or both of the π-bonds of the diene is replaced by a σ-bond, and they also resemble [1,5] sigmatropic rearrangements, in that a σ-bond moves, but from one molecule to another rather than within one molecule. The general cases are illustrated by the drawings **G.9** and **G.10**, in which the R groups are usually hydrogen atoms.

G.9

G.10

Hetero-Diels–Alder reactions are those Diels–Alder reactions in which either the diene is a heterodiene (qv.) or the dienophile is a heterodienophile (qv.), or both components have conjugated systems with an electronegative heteroatom.

Heterodienes are those dienes that have one or more heteroatoms like nitrogen, oxygen, or sulfur in the conjugated system in place of the carbon atoms.

Heterodienophiles are dienophiles (qv.) in which one (or both) of the atoms conjugated together is an electronegative heteroatom like nitrogen, oxygen, or sulfur.

Inverse electron demand is a description applied to cycloadditions in which the component that is normally electron-rich (relatively nucleophilic) is electron-poor (relatively electrophilic), while the other component, normally electron-poor, is electron-rich. A typical example is a Diels–Alder reaction between a diene carrying an electron-withdrawing substituent and a dienophile carrying an electron-donating substituent.

Möbius conjugated system is a cyclic conjugated set of p-orbitals in which the atomic orbitals, instead of being all the same way up around the conjugated system, turn over analogous to the twist in a Möbius strip.

Pericyclic reactions are those reactions with cyclic transition structures in which all bond-forming and bond-breaking takes place in concert, without the formation of an intermediate.

Regioisomers are the two or more possible isomers of a regioselective reaction, such as the isomeric products **G.11** and **G.12** of a Diels–Alder reaction.

major   minor

G.11   G.12

Regioselectivity determines which orientation is adopted in the major pathway by the components of a cycloaddition in which both components are unsymmetrical, as illustrated for a general unsymmetrical Diels–Alder reaction in the formation of more of the regioisomer **G.11** than of **G.12**.

Secondary effects in pericyclic reactions are those effects which determine stereoselectivity, regioselectivity, torqueoselectivity, and reaction rate.

Sigmatropic rearrangements are characterized by the movement of a σ-bond from one position to another, with a concomitant movement of the conjugated systems to accommodate the new bond and fill in the vacancy left behind.

Stereoselective reactions are those reactions in which more of one stereoisomer of the product(s) is produced than of the other stereoisomer(s). The secondary effects (qv.) in pericyclic reactions govern their stereoselectivity, like that seen in the *endo* rule (qv.) for Diels–Alder reactions, among other effects.

Stereospecific reactions are those stereoselective reactions (qv.) in which one stereoisomer of the starting material gives more of one stereoisomer of the product when the other stereoisomer of the starting material gives more of the other stereoisomer of the product. The word is used in contrast with reactions that are *only* stereoselective. The Woodward–Hoffmann rules (qv.) govern the stereospecificity of reactions.

Suprafacial (a) applied to the component of a pericyclic reaction (or drawing of a component of a pericyclic reaction) describes the situation in which two new bonds are formed (or drawn) on the same surface of the conjugated system: **G.13** or **G.14** for a conjugated system of π-bonds, in which the dashed lines represent the direction from which the new bonds are forming, and either **G.15** or **G.16** for a σ bond. The opposite is antarafacial (qv.).

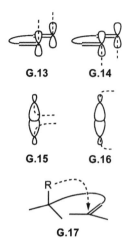

G.13   G.14

G.15   G.16

G.17

Torqueoselectivity in electrocyclizations determines in which sense the allowed stereochemistry is followed in the major pathway; in conrotatory reactions (qv.) it determines whether both bonds rotate clockwise or both anticlockwise, and in disrotatory reactions (qv.) it determines at which end the bond rotates clockwise and at which end anticlockwise.

Woodward–Hoffmann rules take many forms. Originally they were separate rules for each type of reaction, generally one rule for the thermal reaction and its opposite for the corresponding photochemical reaction. The rule might be whether the reaction in question was 'allowed' or 'forbidden'. Alternatively, the rule might state whether the reaction had a particular stereochemistry or its opposite, as in 'conrotatory' (qv.) or 'disrotatory' (qv.). Eventually the rules for thermal reactions became a single rule: a ground-state pericyclic change is symmetry-allowed when the total number of $(4q + 2)_s$ and $(4r)_a$ components is odd. The photochemical rule is then the opposite: a pericyclic change in the first electronically excited state is symmetry-allowed when the total number of $(4q + 2)_s$ and $(4r)_a$ components is even.

X-substituent is an electron-donating substituent like R_2N-, $MeO-$, $AcO-$, or (weakly) alkyl$-$.

Z-substituent is a conjugated electron-withdrawing substituent like $OHC-$, $Ac-$, MeO_2C-, and ArO_2S-.

Suprafacial (b) applied to a sigmatropic rearrangement describes the structural change in which a substituent R attached to a σ bond breaks off from one surface of a conjugated system and forms a σ bond on the same surface at the other end **G.17**.

Synchronous reactions are those concerted reactions (qv.) in which two or more bonds are being made or broken not only at the same time but also to an equal extent in the transition structure.

Index